PRINTED CIRCUIT BOARD

PADS를 이용한 PCB 설계 실무핵심

PCB설계 실무

(PCB: 인쇄회로기판)

윤왕철, 김규희, 함영주 공저

B.C.Info

PREFACE

항상 준비하는 자에게 기회가 온다.

전자공학과를 졸업하고 언젠가 PCB 설계를 해보고 싶었고 내손으로 만들어 보고 싶었다.
봄, 여름, 가을, 겨울이 몇 해 흘렀던가?
이제 교재로 완성되었다.

PCB 설계를 배우고자 하는 전자공학과 관련된 모두에게 도움을 주고 싶다.

1. 전자회로의 지식을 기반으로 하고 PCB 설계를 더 쉽게 배우고자 하는 공학도, 관련학과를 전공한 엔지니어 등 모두가 쉽게 이해하는데 목적을 두고 준비를 했다.

2. PADS를 활용하여 PCB 설계규격 기본개념을 가지고 쉬운 예제를 천천히 따라하면 PCB 설계실무까지 응용하며 접할 수 있다.

3. PCB 설계는 처음 접할 때 처음에서 완성까지 가는 방법을 따라 해 보는 것이 중요하다. PCB 설계는 어려운 것이 아니다.
   산업현장실무에 응용할 수 있도록 교육기회도 부여한다.

교재 편찬을 도와주신 이성훈, 정찬희, 태경, 예진 위원님께 감사드린다.

2021. 10

저자 **윤왕철 · 김규희 · 함영주**

CONTENTS

01 / PADS 9 설치 순서 ········ 7

02 / PCB Artwork 순서도 ········ 16
DATASHEET 이해 ········ 17

03 / Library 이해 ········ 18
Library 이해 (Setup Mode) ········ 22
Library 이해 (Option Mode) ········ 24
Library 이해 (Drafting Toolbar) ········ 25

04 / PADS Layout 시작 ········ 45

05 / PCB Decal 작성 ········ 46
DATASHEET (예제) ········ 66
OR-CAD (PCB Footprint ①) ········ 69
OR-CAD (PCB Footprint ②) ········ 72
OR-CAD (Net list 생성) ········ 73
OR-CAD (BOM 생성) ········ 75

06 / Layout 활용 ········ 76
Layout (Import 활용) ········ 76
Layout (Export 활용) ········ 77
Layout (Find 활용) ········ 78
Layout (Filter 활용) ········ 79
Layout (Nets 활용) ········ 80
Layout (Setup 이해) ········ 81
Layout (PCB Decal Editor) ········ 88
Layout (Pour Manager) ········ 89
Layout (Error 이해) ········ 91
Layout (Basic Scripts) ········ 93
Layout (Option 이해) ········ 94
Layout 이해 (Drafting Toolbar) ········ 102

Layout 이해(Design Toolbar) ········ 117
Layout 이해(ECO Toolbar) ········ 126

07 / Layout 실습(Net list) ········ 135
Layout 실습(Disperse) ········ 137
Layout 실습(Unit 설정) ········ 138
Layout 실습(외형 생성) ········ 139
Layout 실습(부품 배치) ········ 141
Layout 실습(Glued 활용) ········ 144
Layout 실습(Copper) ········ 143
Layout 실습(Copper pour) ········ 144
Layout 실습(Design Toolbar) ········ 145
Layout 실습(ECO) ········ 146

08 / Layout 실습(CAM 이해)(1) ········ 147
Layout 실습(CAM ①) ········ 148
Layout 실습(CAM ②) ········ 151
Layout 실습(CAM ③) ········ 154
Layout 실습(CAM ④) ········ 158
Layout 실습(CAM ⑤) ········ 162
Layout 실습(CAM ⑥) ········ 165
Layout 실습(CAM ⑦) ········ 168
Layout 실습(CAM ⑧) ········ 172

09 / Layout 실습(CAM 이해)(2) ········ 174

10 / HOT KEY ········ 177
PADS 단축키 ········ 177
PADS 명령어 ········ 179
PCB 용어 ········ 181

11 / 실습예제 ········ 190

▌실습회로도 1, 2, 3, 4 ········ 201
▌PARTS DATASHEET ········ 209

01 / PADS 9 설치 순서

01 PADS 설치 프로그램을 실행한다.(SETUP.EXE)

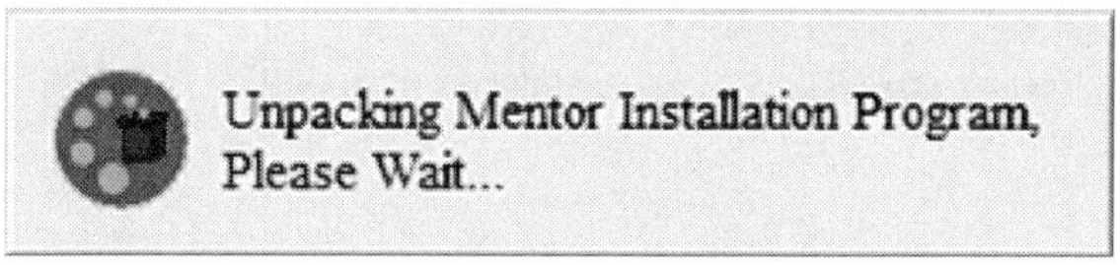

02 Hardware Key 화면에서 Next 버튼을 누른다.

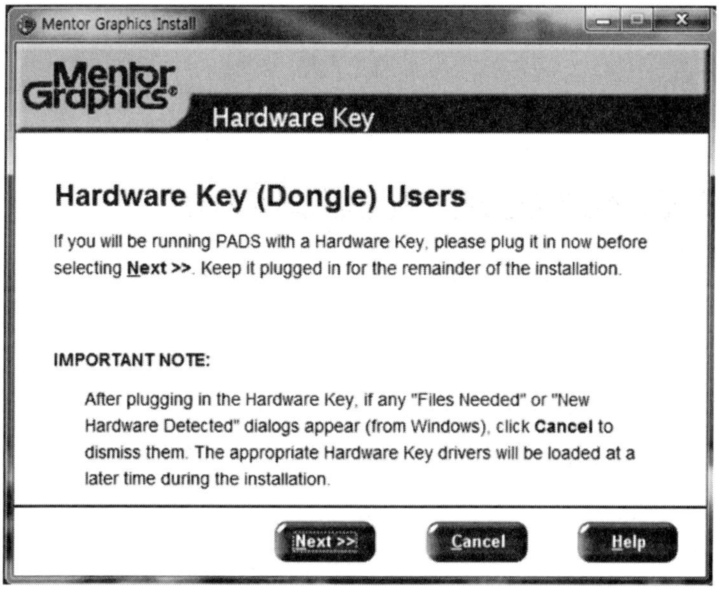

03 Skip을 누른다.

04 Agree를 누른다.

05 Modify를 누른다.

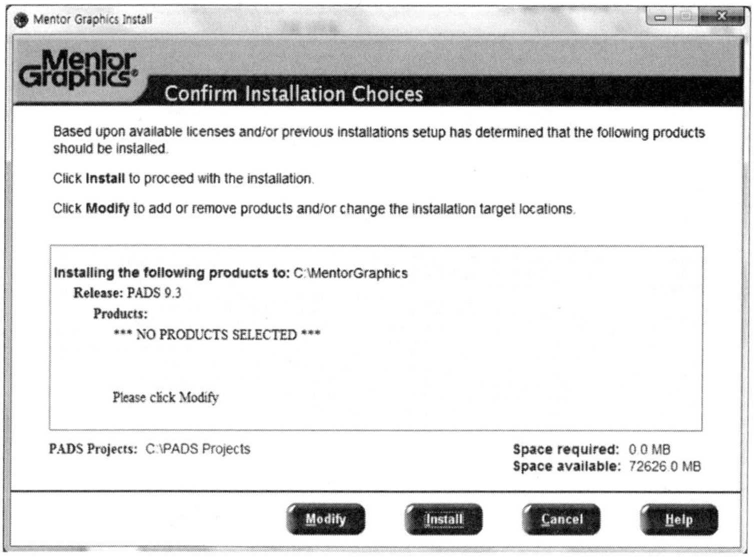

06 Product Selection을 누른다.

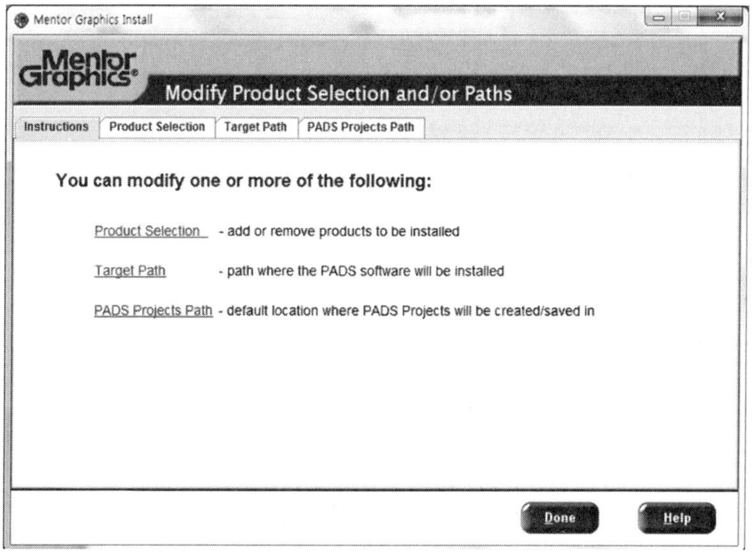

07 설치 프로그램을 선택한 후 Done 버튼을 누른다.

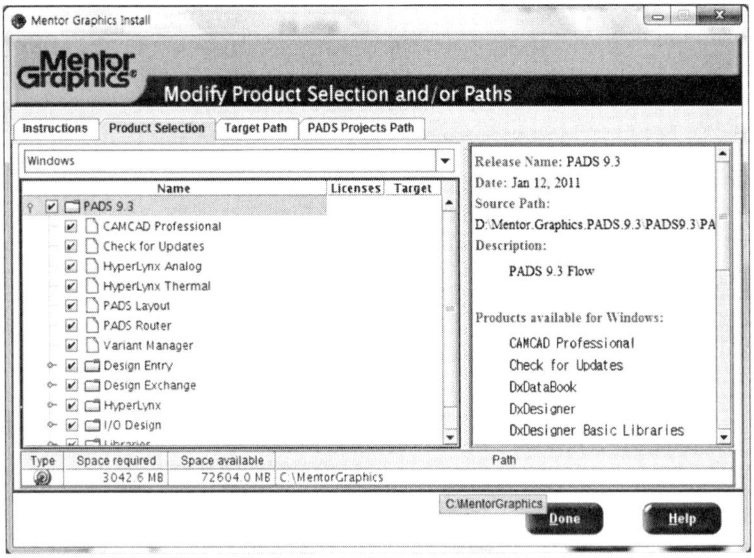

08 Install을 누른다.

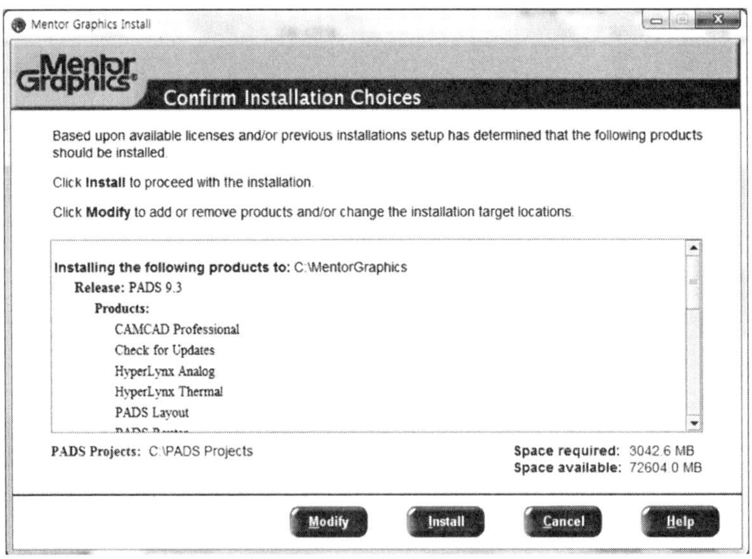

09 설치 중 아래 화면이 나타난다.

10 Next를 누른다.

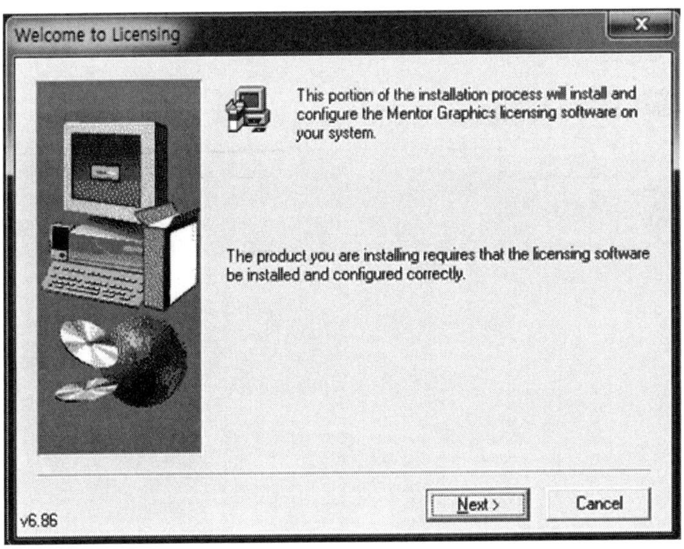

11 Next를 누른다.

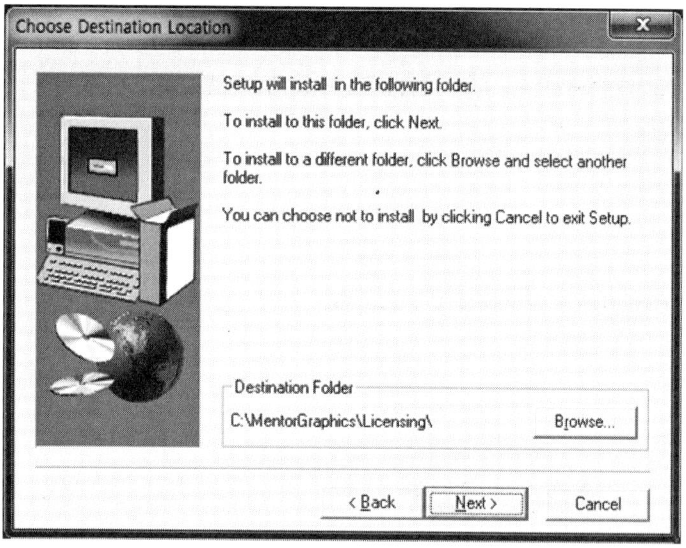

12 설치 중 아래의 화면이 나타난다.

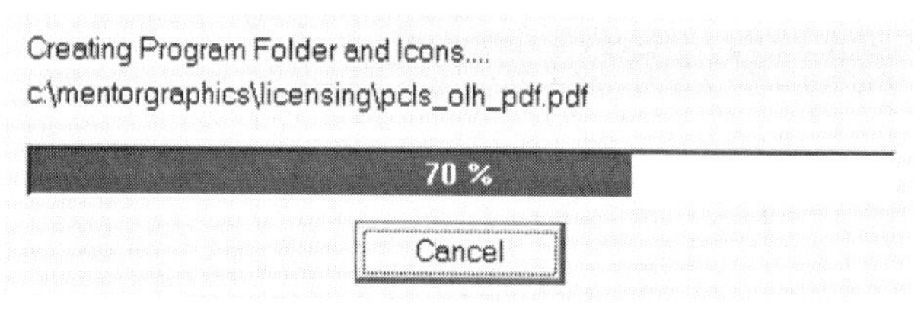

13 Typical을 누른다.

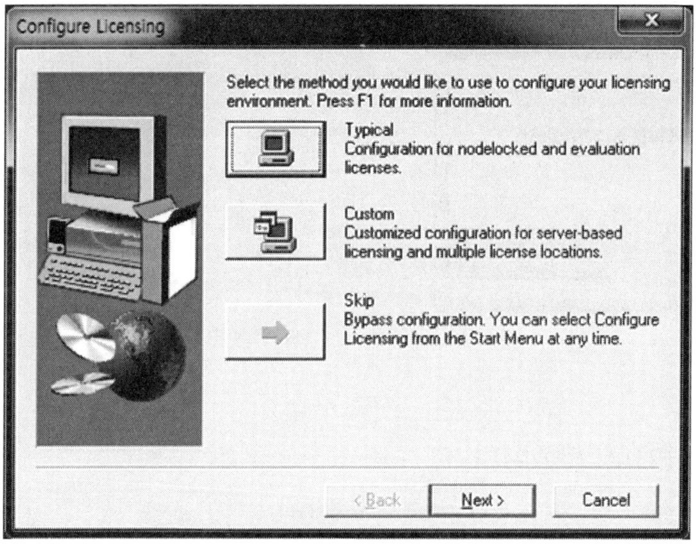

14 Browse를 눌러 License 파일을 선택한 후 Next 버튼을 누른다.

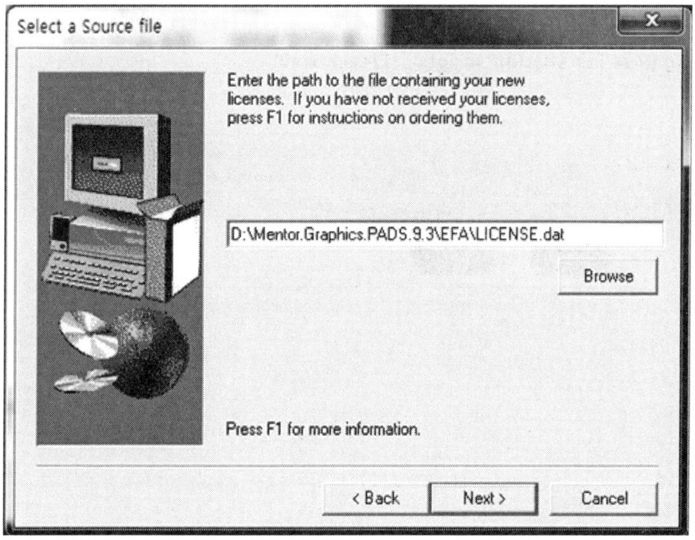

15 License 설정이 완료된다.

16 자동으로 다음단계로 진행된다.

17 At a later time을 선택한 후 Done 버튼을 누른다.

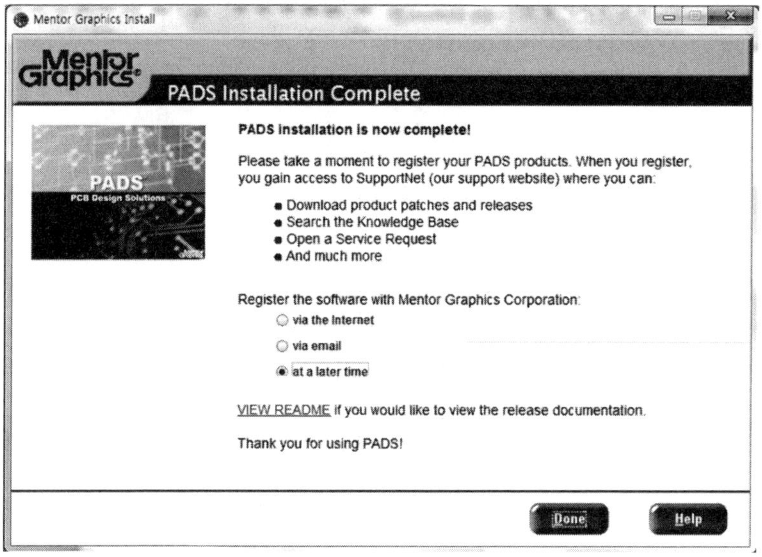

18 프로그램 설치를 완료한다.

02 / PCB Artwork 순서도

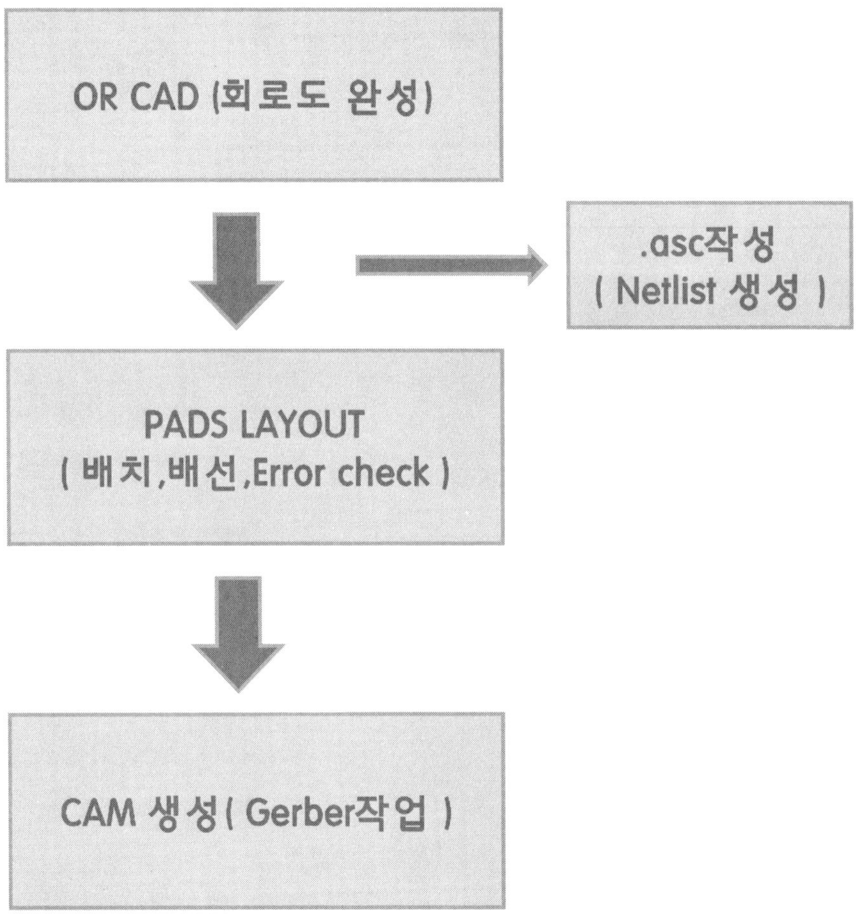

DATASHEET 이해

- **SOT-23-3 (SC-59A)**

Unit: mm

PACKAGE DIMENSIONS

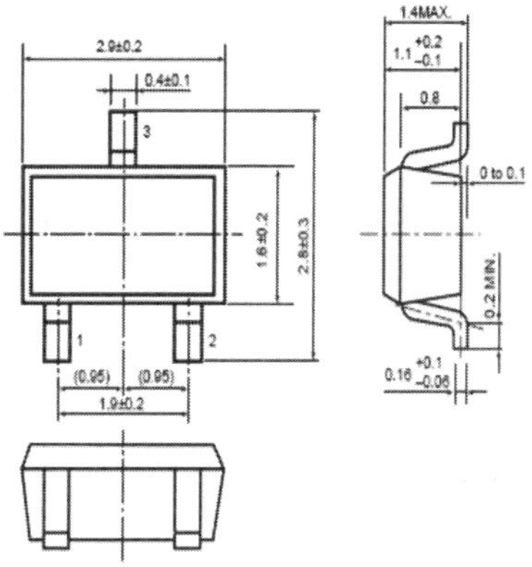

①

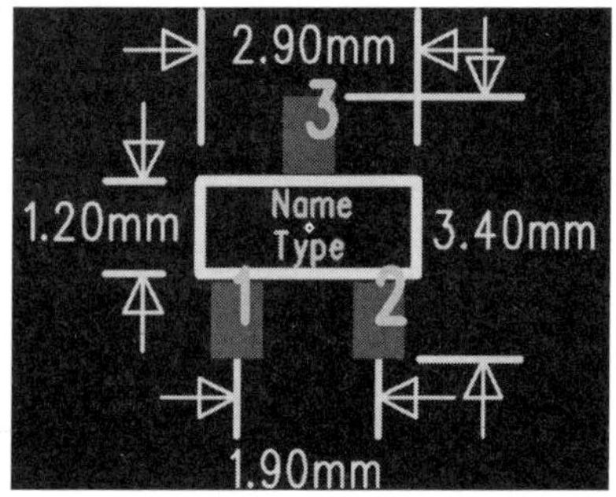

TR의 PCB DECAL 작성

가. DATASHEET를 기준으로 치수를 확인하여 ①과 같이 완성되도록 부품의 외각 SIZE, 핀의 사이즈 및 위치를 적용하여 Library에 등록시킨다.

나. PCB Artwork을 시작하는데 있어 가장 우선으로 해야 하는 과정이 위의 그림과 같이 Decal을 Library에 등록시키는 작업이다.

다. 부품 작성 시 필요한 치수

1. 부품의 외각 size
2. 핀의 정보
   - Pad size, Pin Pitch, Pin의 개수, Pin No.

03 / Library 이해

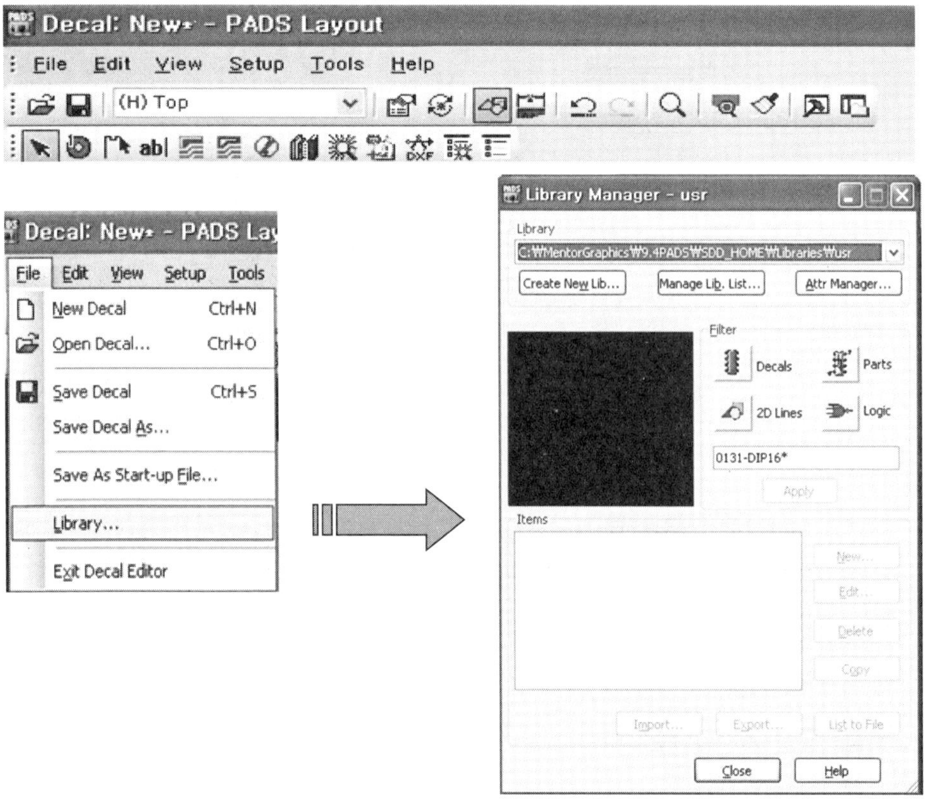

Library 이해하기

가. Library Manager

- 필요한 부품이 Library에 있는지 확인할 수 있다.
- 부품을 삭제하거나 수정하는 작업이 가능하다.
- 여러 개의 Library List를 한눈에 볼 수 있다.

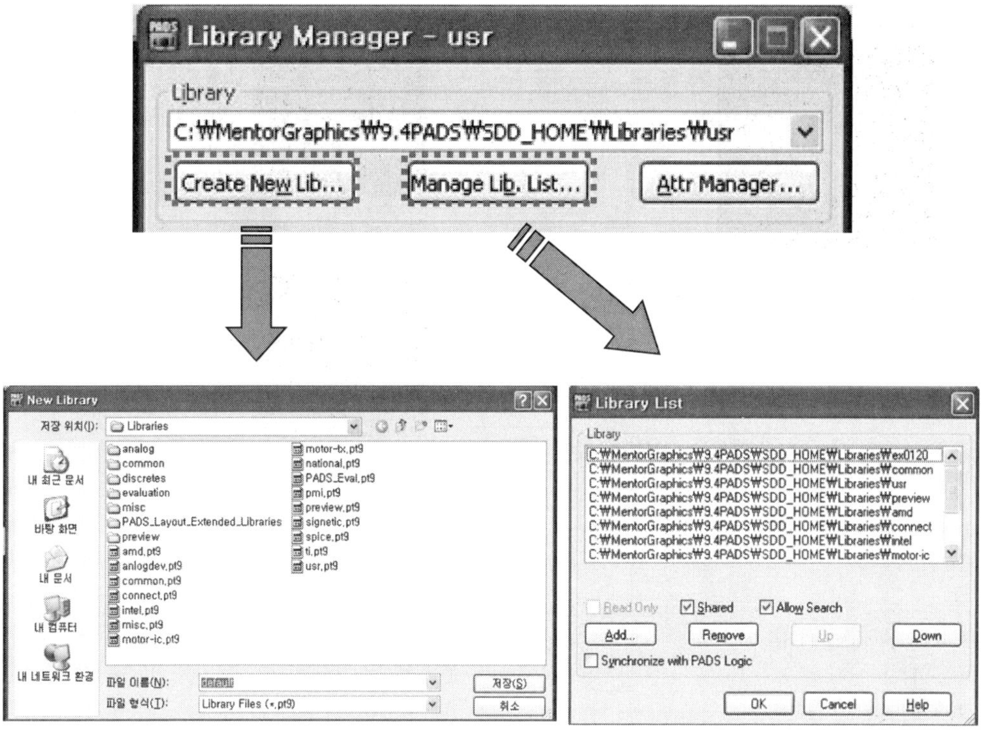

Library 이해하기

가. Create New Library

- 새로운 Library를 추가할 수 있다.

나. Manager Library List

- Library List에서 추가, 삭제 및 순서를 변경할 수 있다.

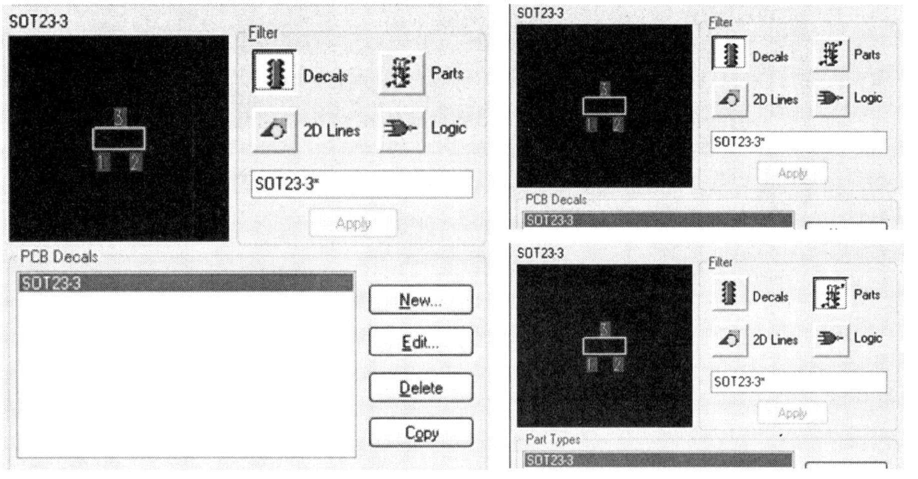

Library 이해하기

가. Filter

- 부품을 Decal과 Part Type으로 나누어 찾을 수 있다.

나. Decal과 Part Type

- Logic에는 Part Type, Layout에는 Decal을 인식한다.
- 선택된 부품을 수정, 삭제하거나 새로운 부품을 만들 수 있다.
- Netlist를 Layout으로 열 때 Part Type이 없을 경우 부품이 불러지지 않아 Error가 발생한다.

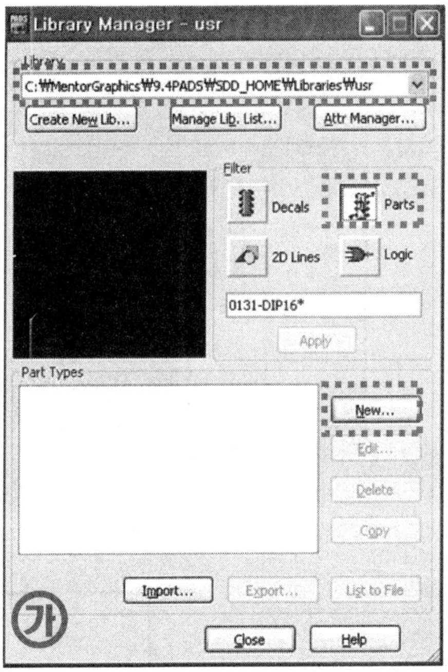

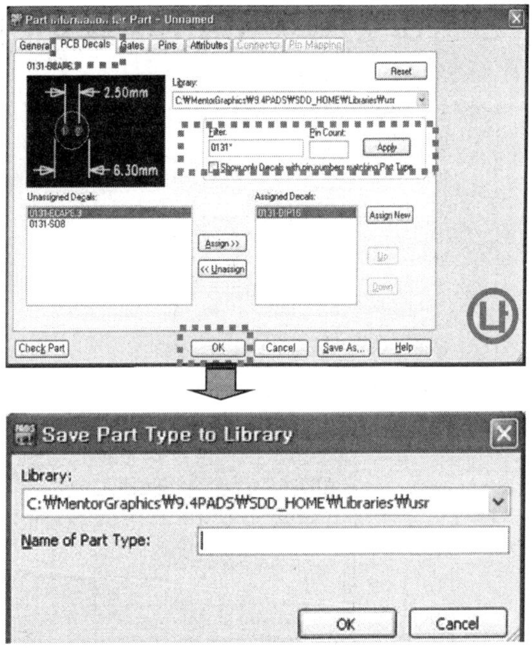

Netlist를 Layout으로 열 때 Part Type이 없을 경우

가. Library Manager - Filter (Parts) – XXX* – Apply – New

- Part Type에 부품이 없는지 확인 후에 새로 추가한다.

나. Part Information For Part – PCB Decal – Filter – XXX* – Apply
– Unassigned Decals: – XXX 선택 – Assign – Assigned Decals:이동
– OK – Save Part Type to Library - XXX 저장 – OK

- Part Type을 새로 Library에 저장하여 사용한다.

Library 이해 (Setup Mode)

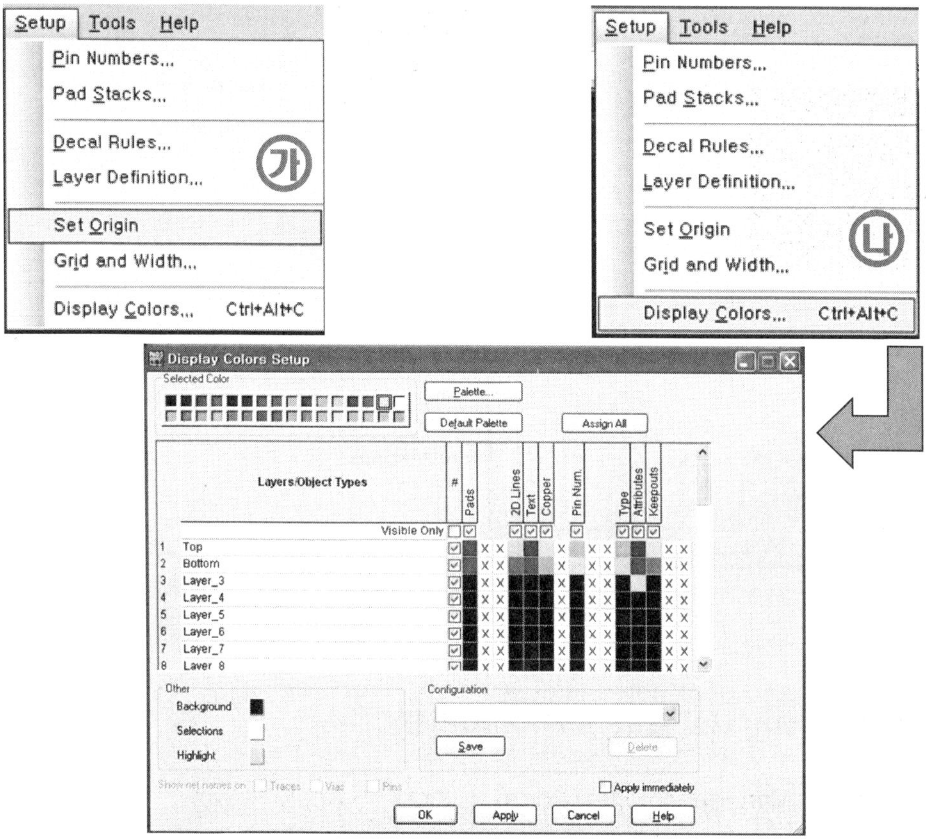

Setup Mode

가. Setup – Set Origin

- 부품 작성시 Origin을 변경할 수 있다.

나. Setup – Display Colors – Display Colors Setup

- 부품의 Top과 Bottom의 각 Item을 색깔로 구분할 수 있다.

Library 이해 (Setup Mode)

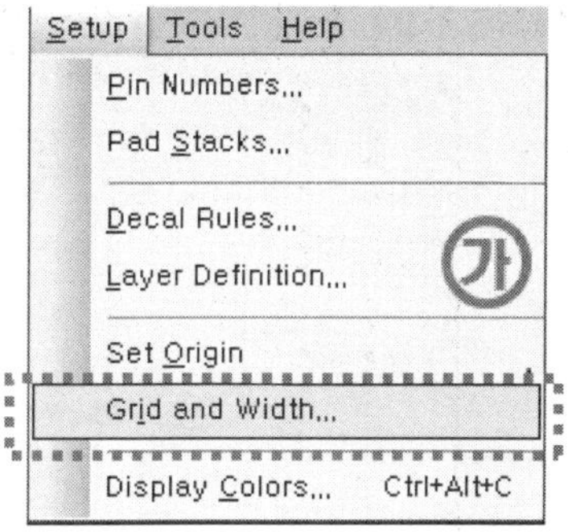

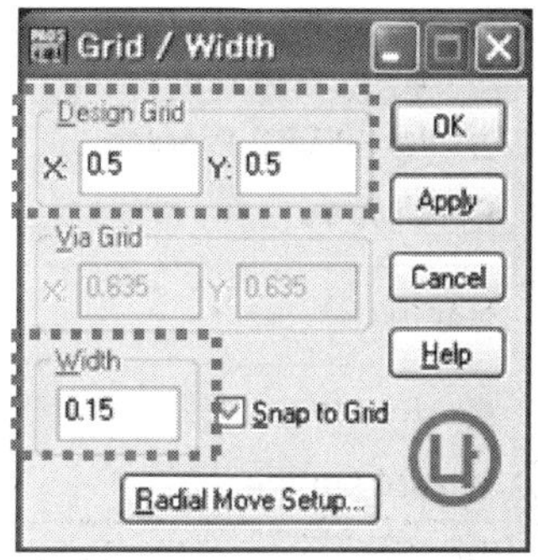

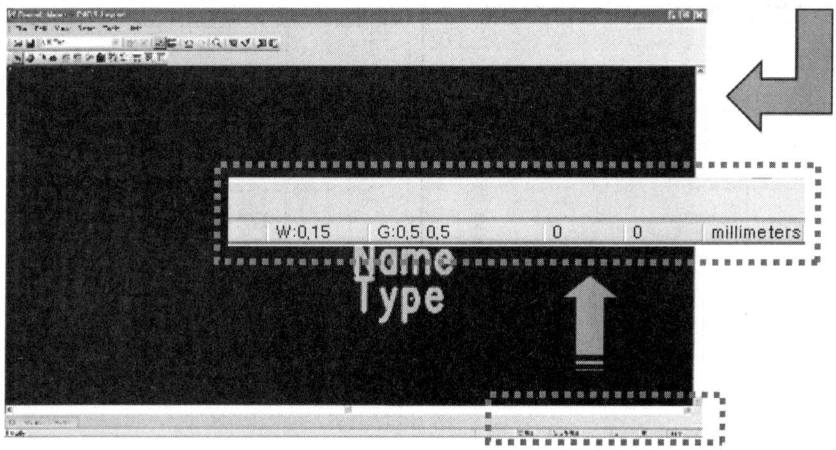

Setup Mode

가. Setup -Grid and Width

- 부품 작성시 Grid 와 Width를 변경할 수 있다.

나. Grid / Width -Design Grid(0.5/0.5) -Width(0.15)

- Cursor의 최소단위(Grid)와 2D Line의 두께를 설정할 수 있다.

Library 이해 (Option Mode)

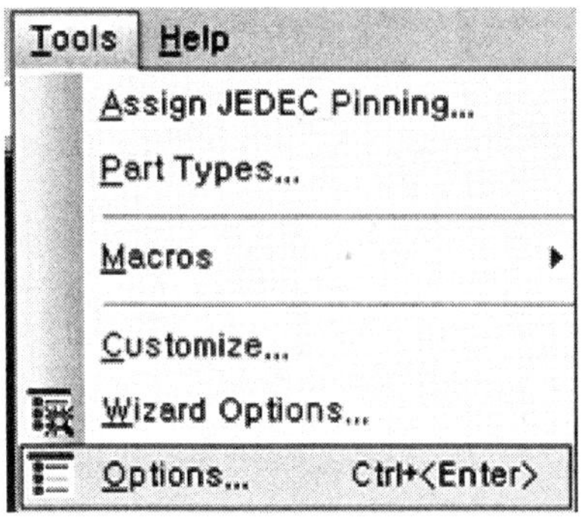

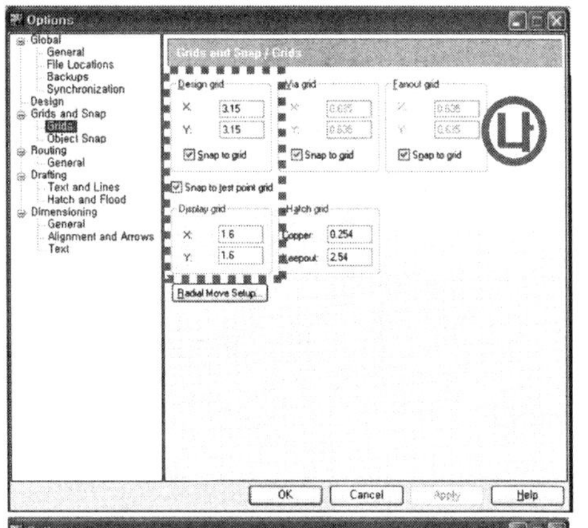

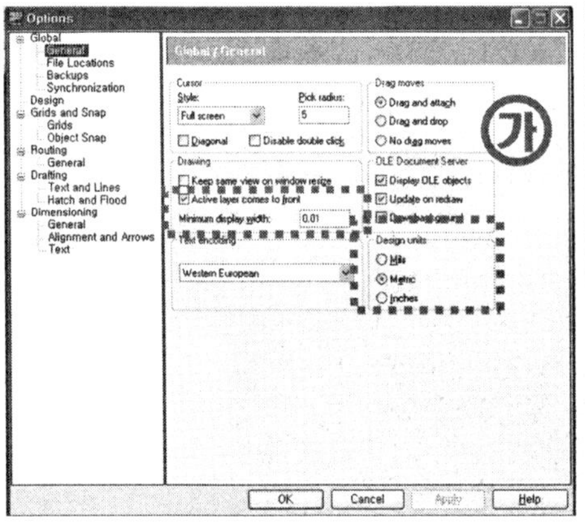

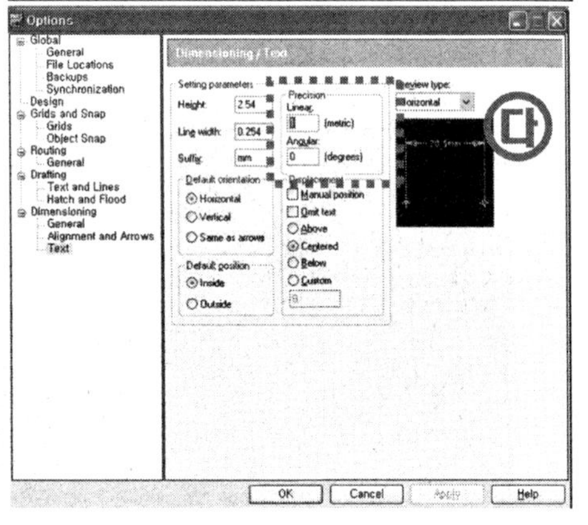

Option Mode

가. Tools – Option – Global – General

- 단위설정 및 최소 두께 설정을 할 수 있다.

나. Tools – Option – Grids and Snap – Grids

- Design Grid (커서 최소단위), Display Grid (화면상의 포인트의 간격)를 설정한다.

다. Tools – Option – dimensioning – Text – Precision – Linear – 2

- 거리치수 측정 시 단위를 소수점 2자리로 표시한다.

Library 이해 (Drafting Toolbar)

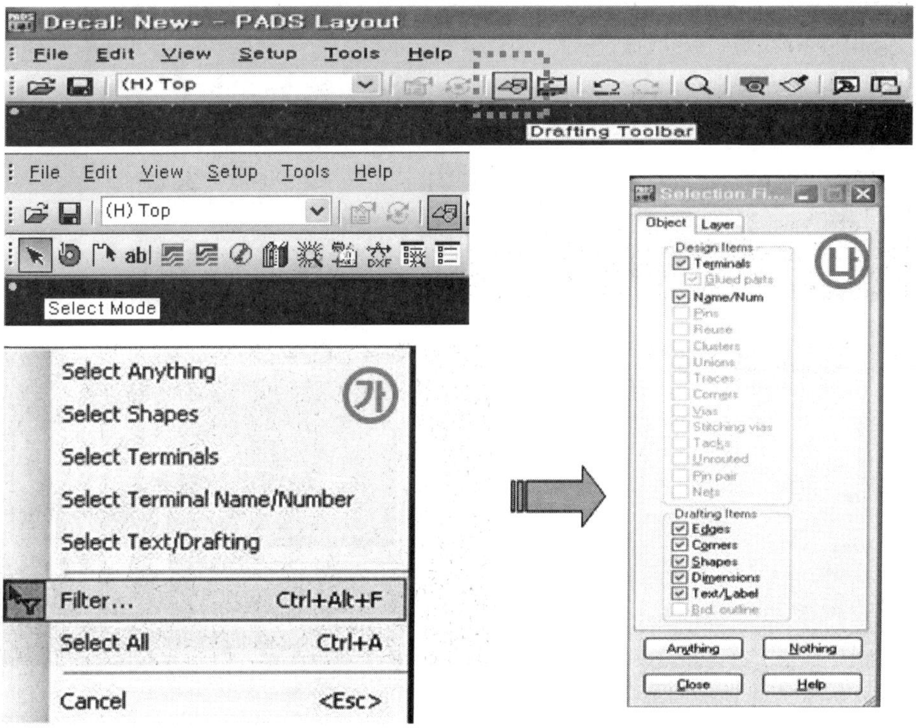

Select Mode

가. Drafting Toolbar – Select -마우스(우측)　*단. 선택된 Item 없을 경우

- 자주 쓰이는 Items을 선택할 수 있다.

나. Filter – Anything or Nothing

- 세분화된 Item을 선택할 수 있다.
- Design Items (Terminal=Pin, Name/Num=Pin Name/Pin No.)
- Drafting Items (Edges, Corner, shape=2D Line, Dimension=거리 측정 치수, Text/Label=부품 Type/Name)

Library 이해(Drafting Toolbar)

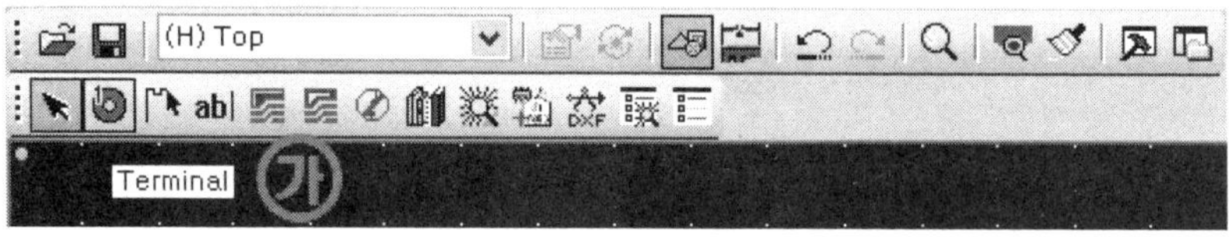

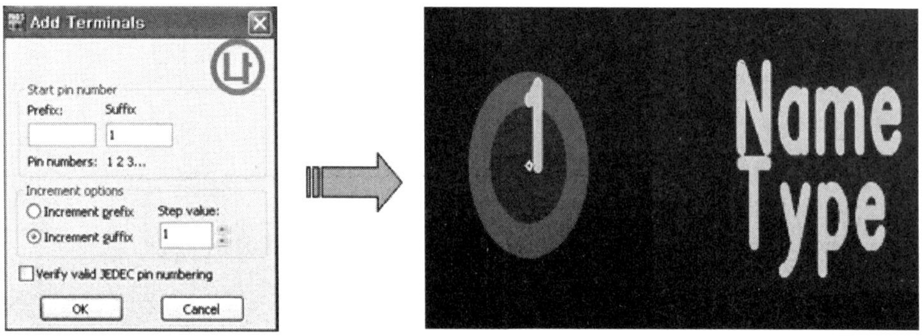

Terminal Mode

가. Drafting Toolbar – Terminal

- Pin, Pin number를 생성한다.

나. Drafting Toolbar – Terminal -Add Terminal – Start pin number – Suffix(Pin number) -OK – 위치 선정 -마우스 (좌측) -Pin 생성

- Pin number를 선택하고 원하는 위치에 Pin을 생성한다.

Library 이해 (Drafting Toolbar)

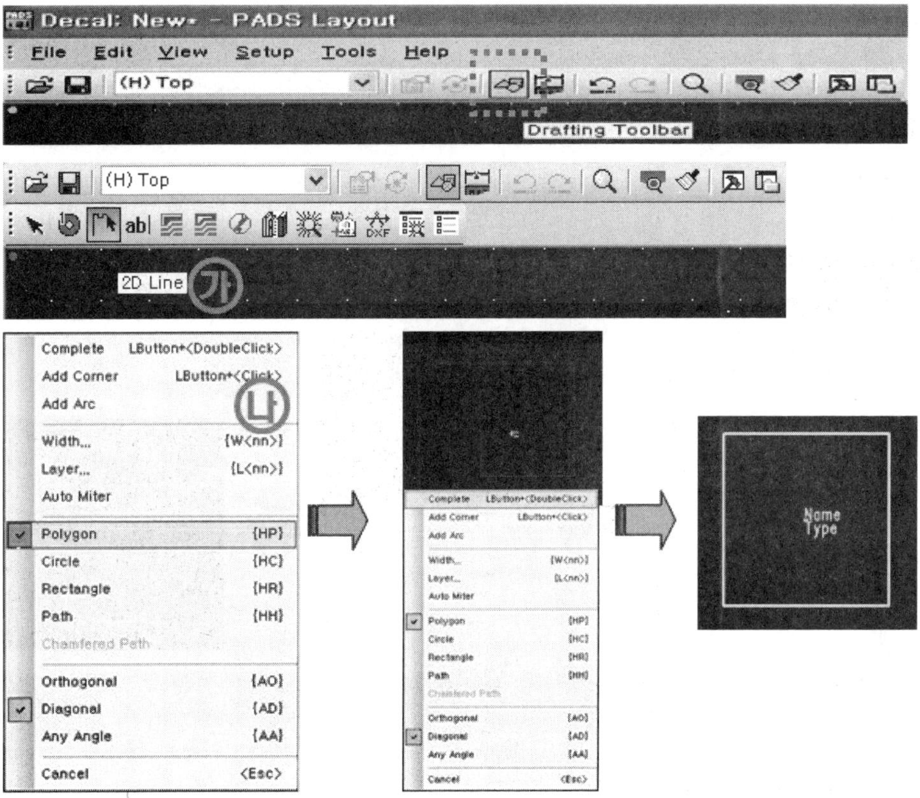

2D Line Mode (Polygon)

가. Drafting Toolbar -2D Line

- 부품의 외형이나 Pin Number 표시를 생성할 수 있다.

나. Drafting Toolbar – 2D Line -마우스(우측) -Polygon -도형생성 -마우스(우측) – Complete

- Polygon : 선으로 그린 도형을 box로 완성할 수 있다.

Library 이해(Drafting Toolbar)

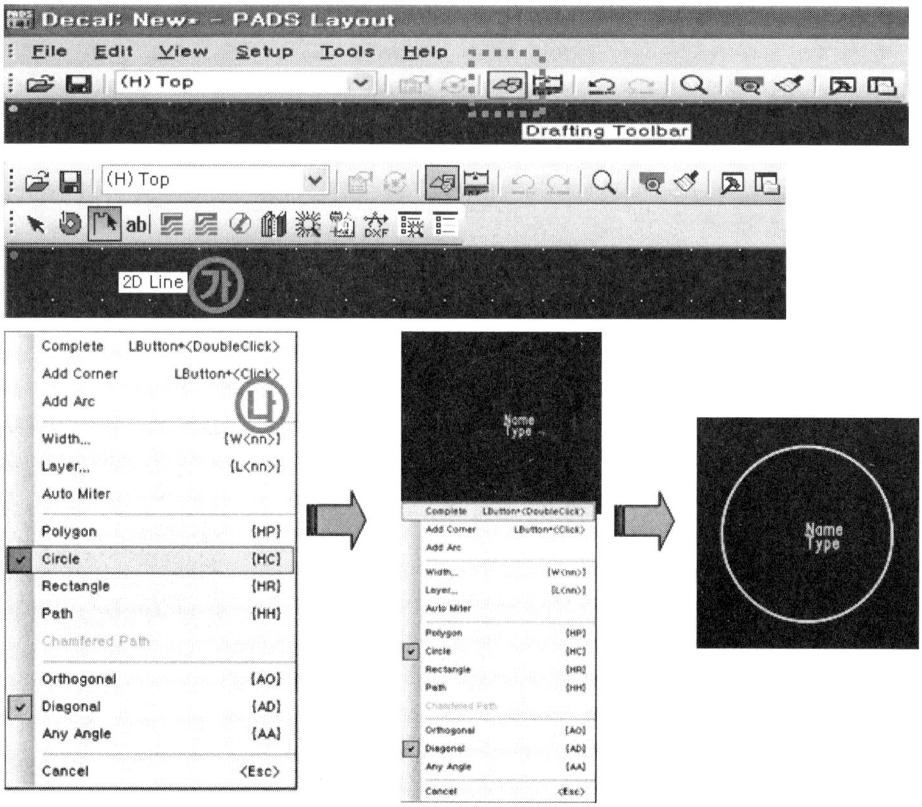

2D Line Mode (Circle)

가. Drafting Toolbar -2D Line

- 부품의 외형이나 Pin Number 표시를 생성할 수 있다.

나. Drafting Toolbar – 2D Line -마우스(우측) -Circle -도형생성 – 마우스(우측) – Complete

- Circle : 원형의 도형을 완성할 수 있다

Library 이해 (Drafting Toolbar)

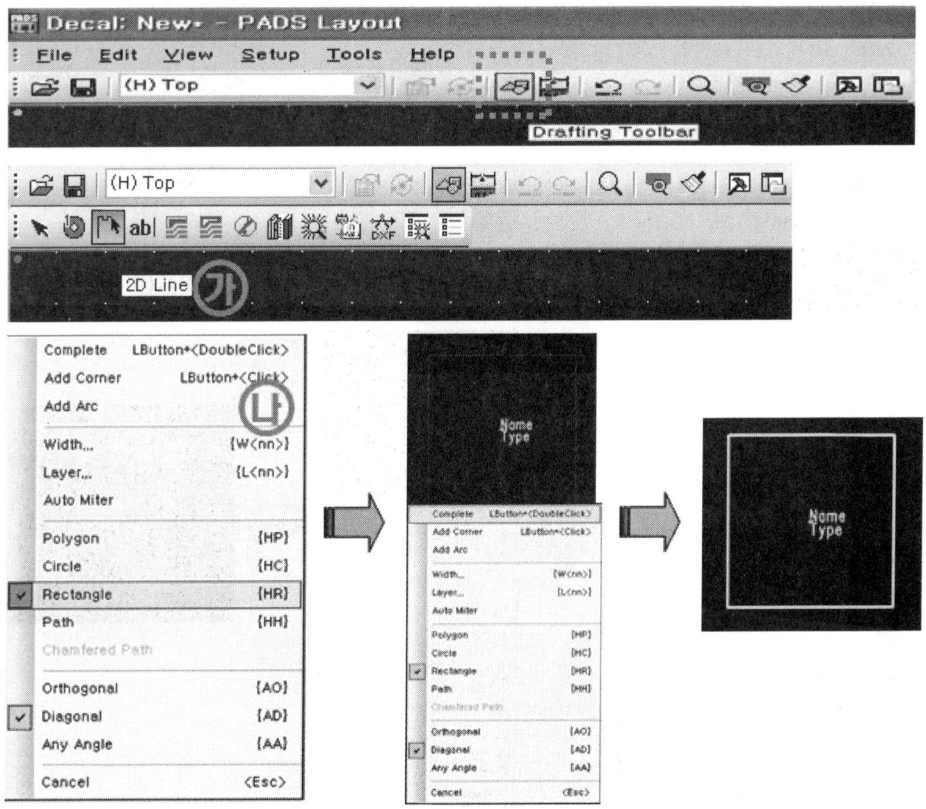

2D Line Mode (Rectangle)

가. Drafting Toolbar -2D Line

- 부품의 외형이나 Pin Number 표시를 생성할 수 있다.

나. Drafting Toolbar – 2D Line -마우스(우측) -Rectangle -도형생성 -마우스(우측) – Complete

- Rectangle : Box의 도형을 완성할 수 있다.

Library 이해(Drafting Toolbar)

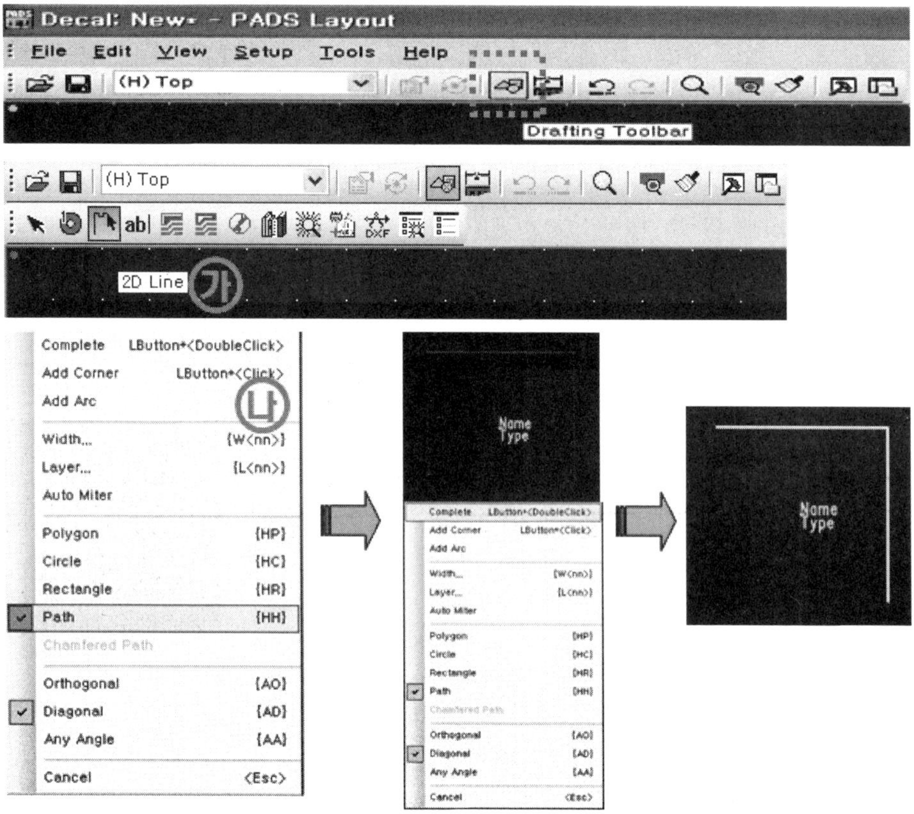

2D Line Mode (Path)

가. Drafting Toolbar -2D Line

- 부품의 외형이나 Pin Number 표시를 생성할 수 있다.

나. Drafting Toolbar – 2D Line -마우스(우측) -Path -도형생성 -마우스(우측) – Complete

- Path : 선의 도형을 완성할 수 있다.

Library 이해 (Drafting Toolbar)

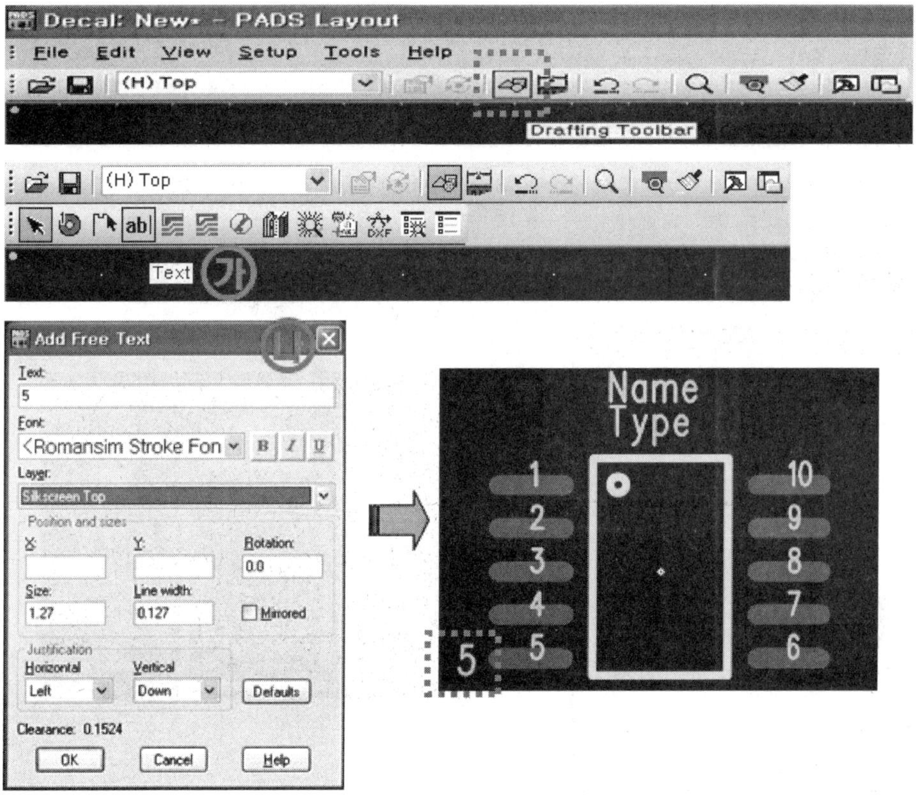

Text Mode (TOP Layer 생성시)

가. Drafting Toolbar -Text

- 부품의 Pin number, 부품 이름을 생성할 수 있다.

나. Drafting Toolbar – Text -Add Free Text -Text(숫자) -Layer(Silkscreen Top) -Size(Text크기) -Line width(Text 두께) – OK

- Top Layer에 Text의 레이어, 크기, 두께를 조절할 수 있다.

Library 이해(Drafting Toolbar)

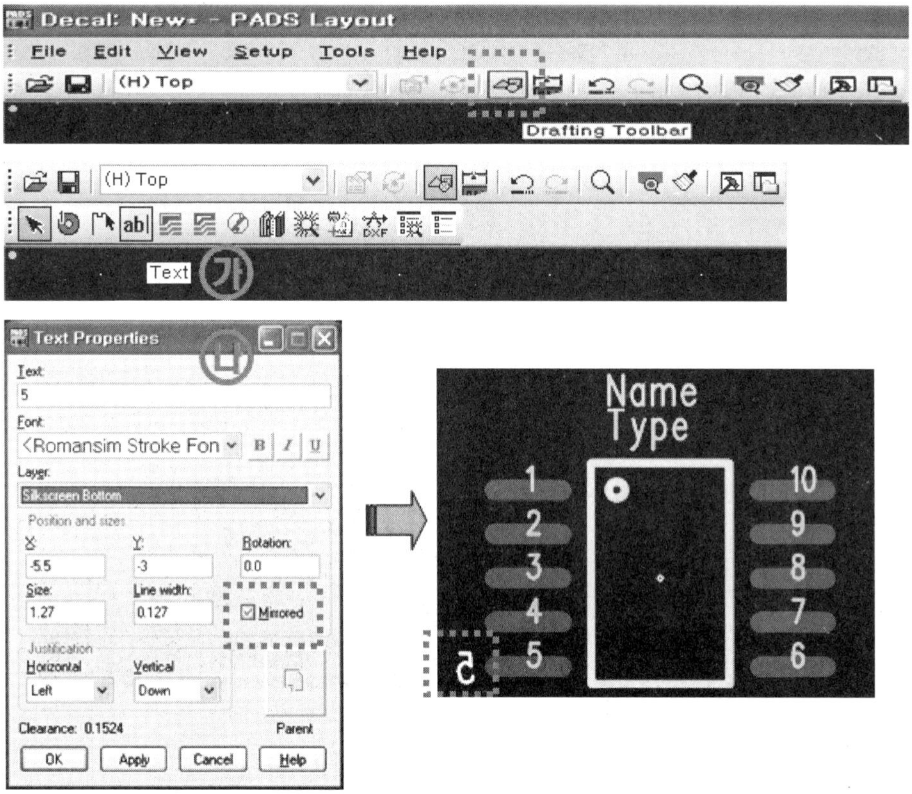

Text Mode (BOTTOM Layer 생성시)

가. Drafting Toolbar -Text

- 부품의 Pin number나 PCB Board name을 생성할 수 있다.

나. Drafting Toolbar - Text -Add Free Text -Text(숫자) -Layer(Silkscreen Bottom) -Size(Text크기) -Line width(Text 두께) -Mirrored 확인 - OK

- Bottom Layer에 Text의 레이어, 크기, 두께를 조절할 수 있다.

Library 이해 (Drafting Toolbar)

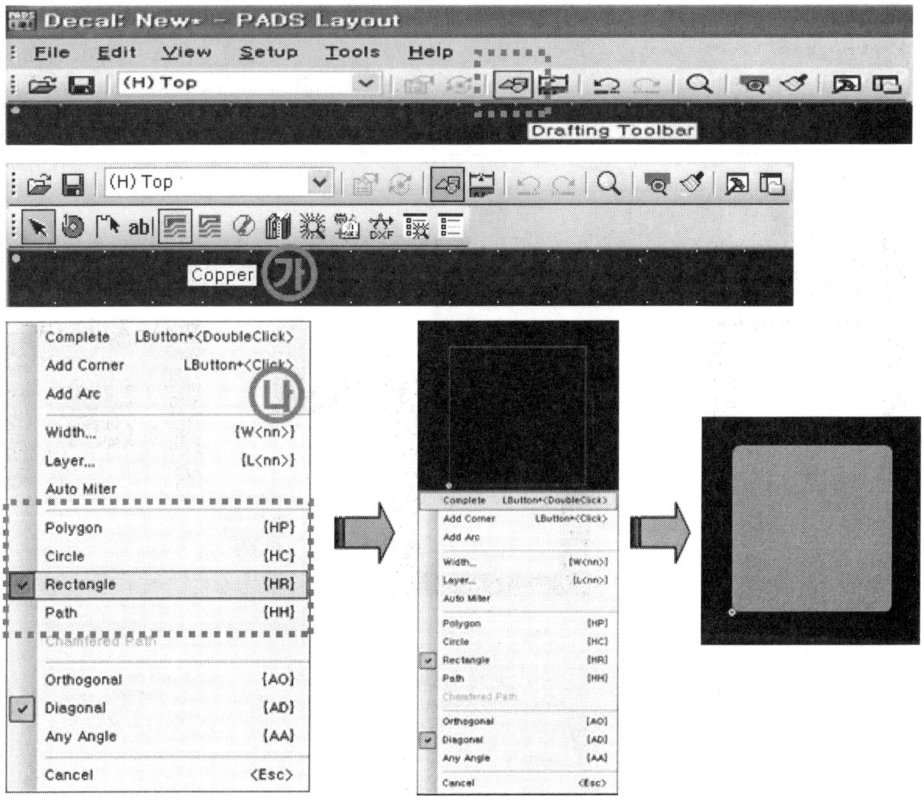

Copper Mode

가. Drafting Toolbar -Copper

- Filled된 도형을 그릴 수 있다. (Polygon, Circle, Rectangle, Path)

나. Drafting Toolbar – Copper -마우스(우측) -Rectangle -도형생성 -마우스(우측) -Complete

- 부품의 외형의 실크추가에 활용할 수 있다.

Library 이해(Drafting Toolbar)

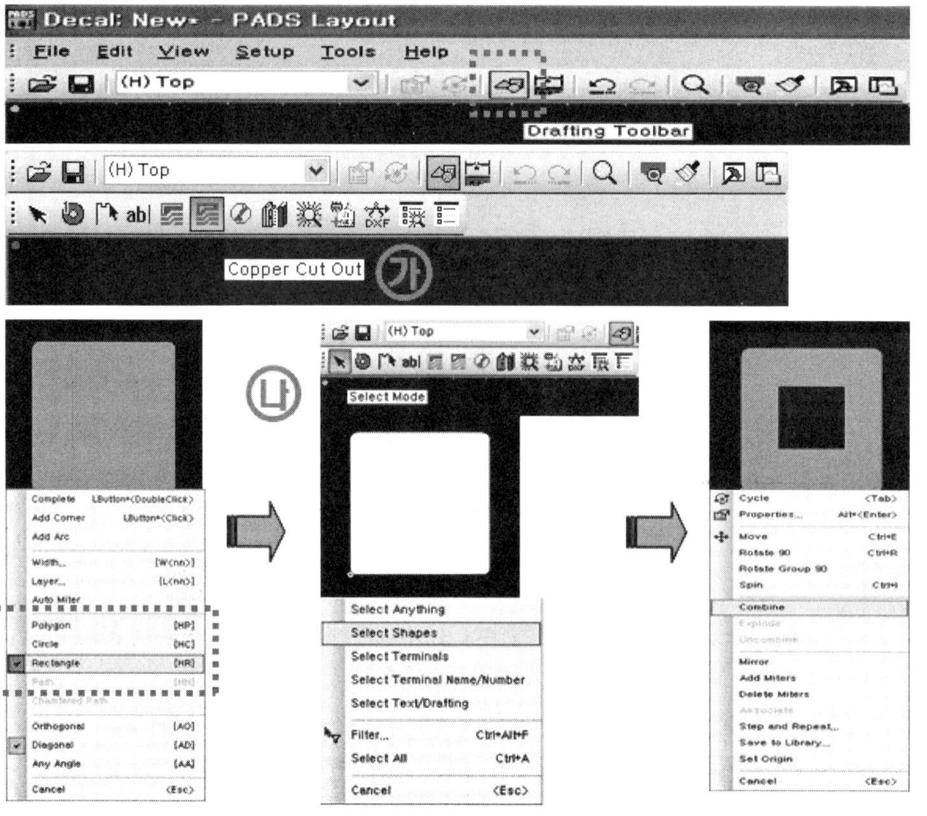

Copper Cut Out Mode

가. Drafting Toolbar -Copper Cut Out

- Filled된 도형을 잘라 낼 수 있다. (Polygon, Circle, Rectangle)

나. Drafting Toolbar – Copper Cut Out-마우스(우측) -Rectangle -도형생성 - Select Mode -마우스(우측) -Select Shapes -전체도형 선택 – 마우스(우측) - Combine

- 부품의 외형의 실크추가에 활용할 수 있다.

Library 이해 (Drafting Toolbar)

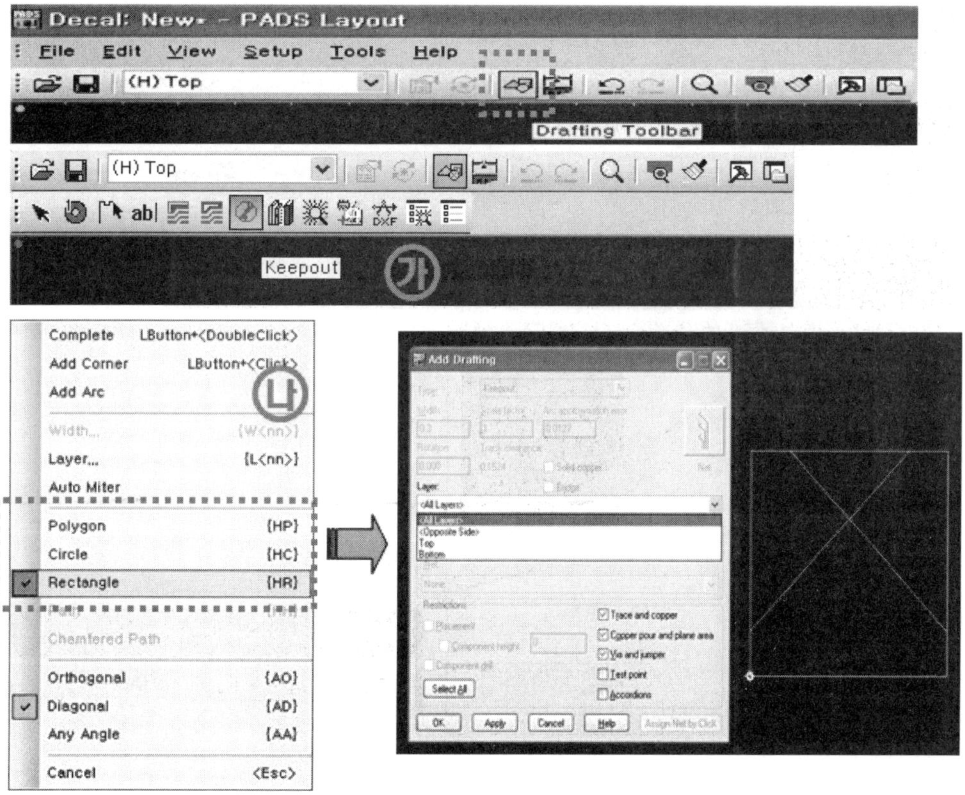

Keep Out Mode

가. Drafting Toolbar -Keep out

- 부품의 금지구역 대를 설정할 수 있다. (Polygon, Circle, Rectangle)

나. Drafting Toolbar – Keep out -마우스(우측) -Rectangle -도형생성 -Add Drafting -Layer선택 (All, Top, Bottom) -Restrictions (5가지 중 복수 선택가능) – OK

- 부품의 금지구역 대를 설정하여 Layer별로 5가지의 Item을 금지시킬 수 있다.

Library 이해(Drafting Toolbar)

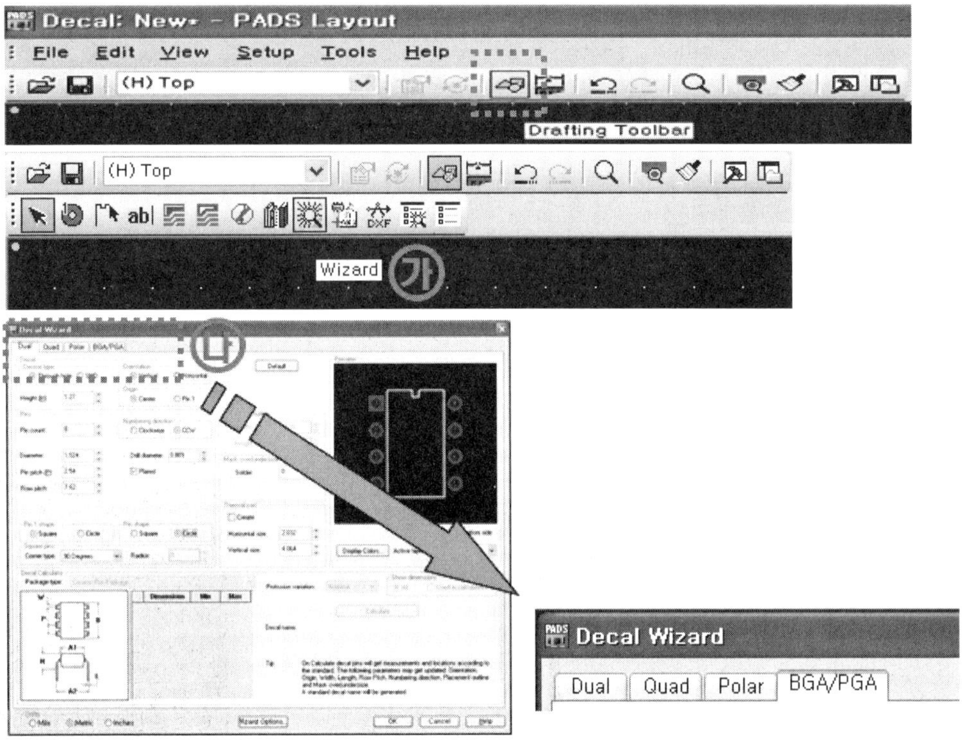

Wizard Mode

가. Drafting Toolbar -Wizard – Decal Wizard -4가지 타입 중 선택

- 부품의 타입 별로 생성이 가능하다. (Dual, Quad, Polar, BGA/PGA)
- Dual : Dip, SOIC, SOJ Type IC 를 생성할 수 있다.
- Quad : QFP, PLCC Type의 IC를 생성할 수 있다.
- Polar : 방사형 부품을 생성할 수 있다.
- BGA/PGA : BGA Type IC를 생성할 수 있다.

Library 이해 (Drafting Toolbar)

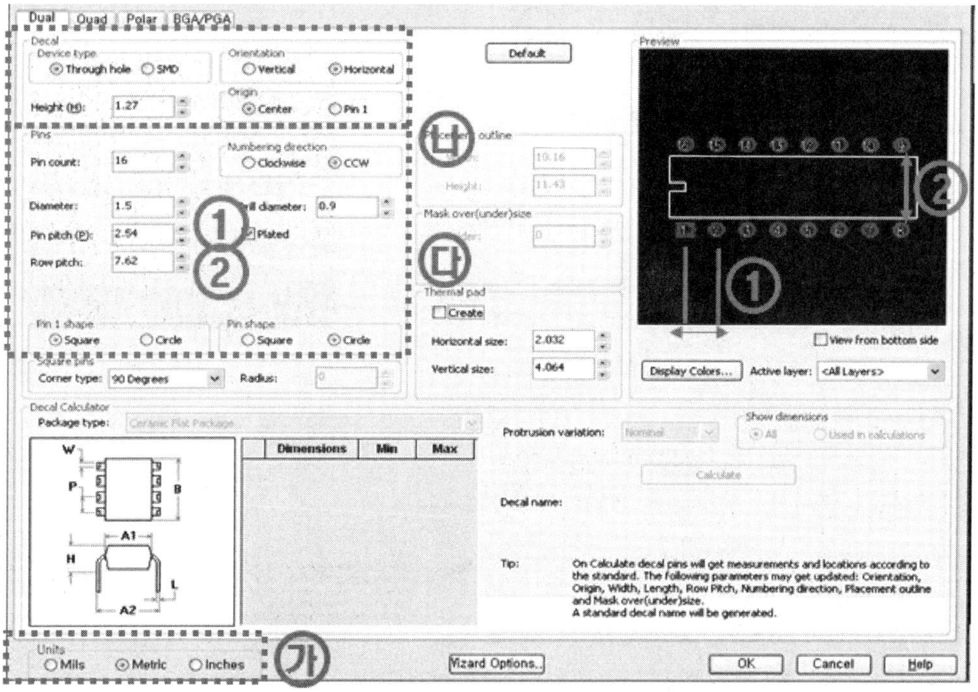

Wizard Mode (Dual-DIP Type)

가. Wizard -Decal Wizard – Dual – Units

- 3가지의 Units 중 단위를 설정할 수 있다. (Mil, Metric, Inches)

나. Wizard -Decal Wizard – Dual -Decal

- Device type (DIP/SMD), Orientation(부품의 방향), Origin(부품의 원점)

다. Wizard -Decal Wizard – Dual -Pins

- Pin Count (Pin수), Numbering direction (Pin 번호방향), Diameter / Drill (Pin의 Land size/Hole), ①Pin Pitch (Pin의 간격), ②Row Pitch(2개의 열 간격), Pin 1 Shape (1번 Pin의 모양), Pin Shape (1번 Pin을 제외한 Pin의 모양)선택

Library 이해(Drafting Toolbar)

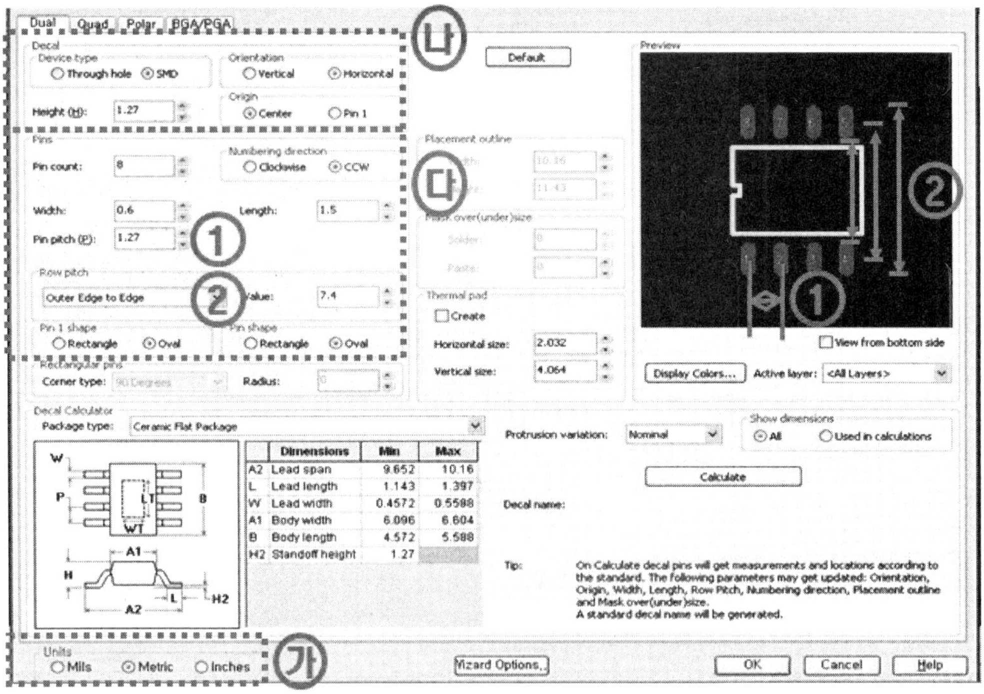

Wizard Mode (Dual-SMD Type)

가. Wizard -Decal Wizard – Dual – Units

- 3가지의 Units 중 단위를 설정할 수 있다. (Mil, Metric, Inches)

나. Wizard -Decal Wizard – Dual -Decal

- Device type (DIP/SMD), Orientation(부품의 방향), Origin(부품의 원점)

다. Wizard -Decal Wizard – Dual -Pins

- Pin Count (Pin수), Numbering direction (Pin 번호방향), Width / Length (Pin의 두께/길이), ① Pin Pitch (Pin 간격), ② Row Pitch (Center/Inner/Outer Edge중 선택, value(열의 간격)), Pin 1 Shape , Pin Shape 선택

Library 이해(Drafting Toolbar)

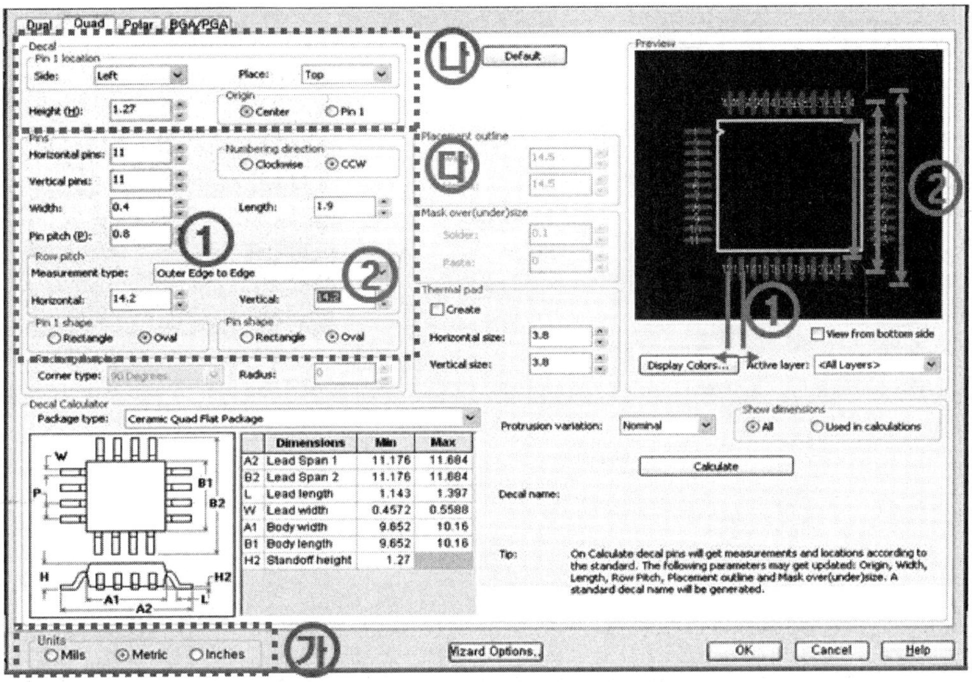

Wizard Mode (Quad IC)

가. Wizard -Decal Wizard – Dual – Units

- 3가지의 Units 중 단위를 설정할 수 있다. (Mil, Metric, Inches)

나. Wizard -Decal Wizard – Dual -Decal

- Pin 1 Location Side(4열 중 선택) / Place(선택 열 중 위치선택), Origin (원점)

다. Wizard -Decal Wizard – Quad -Pins

- Horizontal Pin (가로Pin수), Vertical Pin (세로Pin수), Numbering direction(Pin번호방향), Width / Length (Pin의 두께/길이), ① Pin Pitch (Pin의 간격), ② Row Pitch(Center/Inner/Outer Edge중 선택, Horizontal(가로 열 간격) Vertical(세로 열 간격), Pin 1 Shape, Pin Shape 선택

Library 이해(Drafting Toolbar)

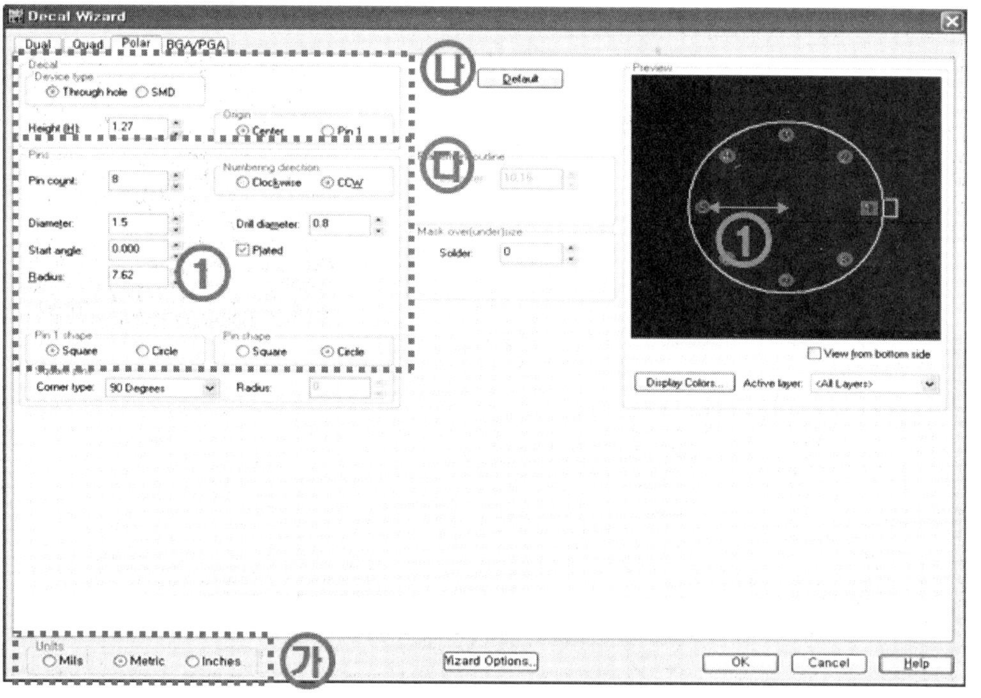

Wizard Mode (Polar)

가. Drafting Toolbar -Wizard -Polar – Units

- 3가지의 Units 중 단위를 설정할 수 있다. (Mil, Metric, Inches)

나. Drafting Toolbar -Wizard -Polar -Decal

- 부품 Device type (DIP/SMD), Origin(부품의 원점)

다. Drafting Toolbar -Wizard -Polar -Decal

- Pin Count (pin수), Numbering direction (Pin번호방향), Diameter / Drill (Pin의 Land size/Hole), Start Angle(부품각도) ①Radius(반지름 거리), Pin 1 Shape, Pin Shape 선택

Library 이해 (Drafting Toolbar)

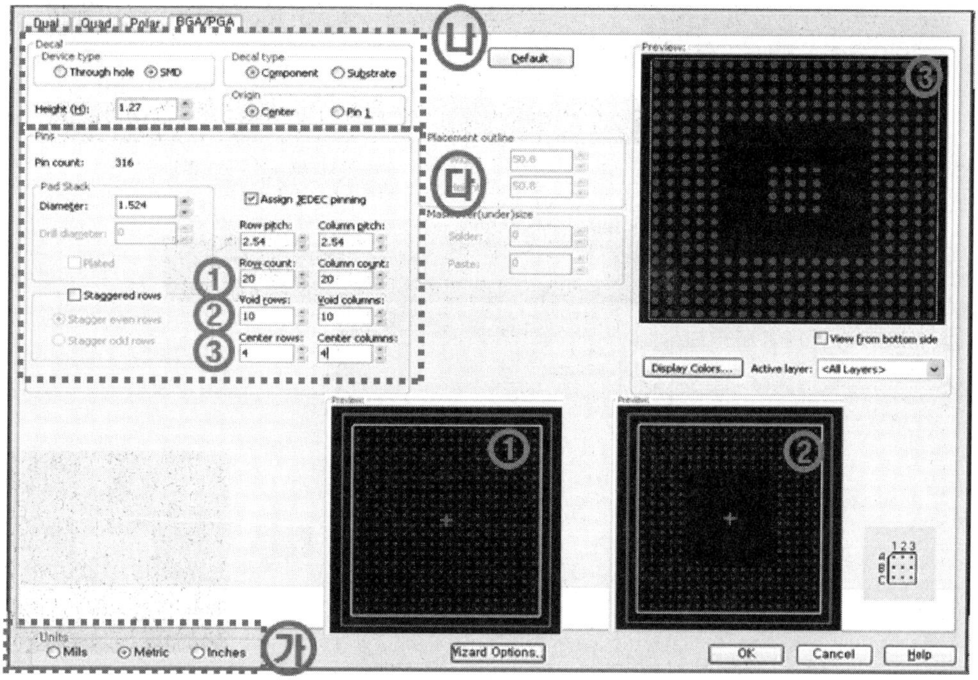

Wizard Mode (BGA/PGA IC)

가. Wizard -Decal Wizard -BGA/PGA -Units

- 3가지의 Units 중 단위를 설정할 수 있다. (Mil, Metric, Inches)

나. Wizard -Decal Wizard -BGA/PGA -Decal

- Device type (DIP/SMD), Decal Type(1번 Pin 방향), Origin(부품의 원점)

다. Wizard -Decal Wizard -BGA/PGA -Pins

- Diameter (Pin의 Land size/Hole), Pin Pitch (Pin간격: Row / Column), ① Count (Pin수 : Row / Column), ② Void (삭제 Pin수 : Row / Column), ③ Center (추가 Pin수 : Row / Column)

Library 이해 (Drafting Toolbar)

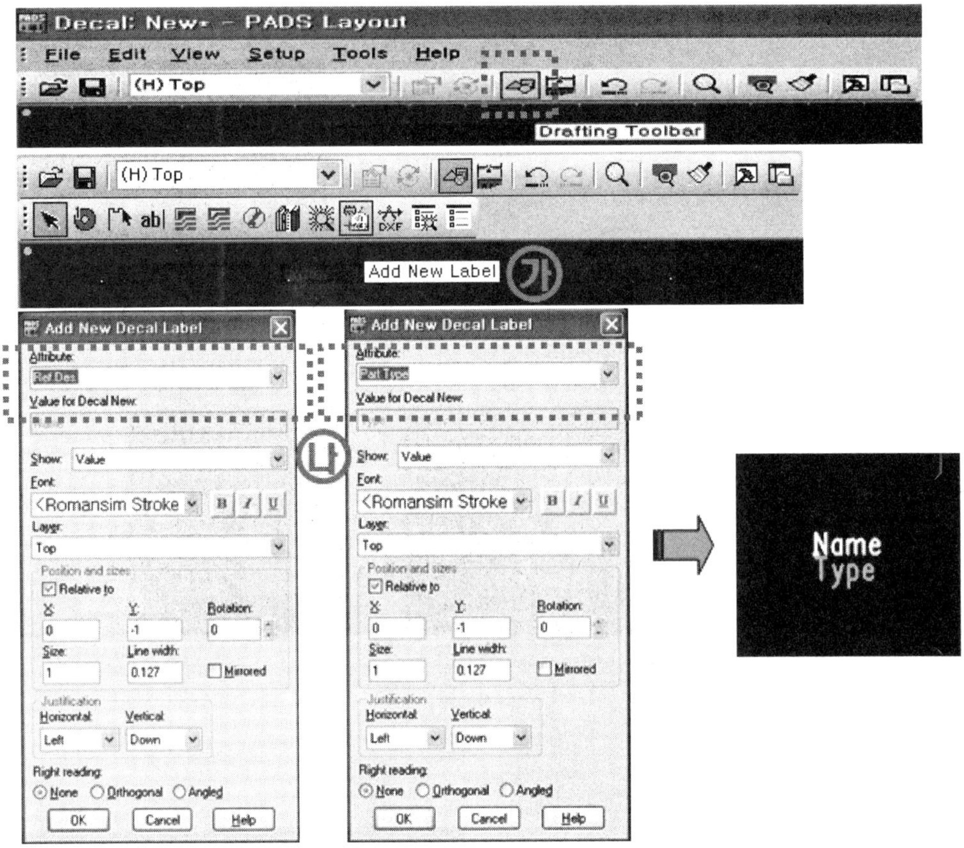

Add New Label Mode

가. Drafting Toolbar -Add New Label

- 부품의 Reference, Part Type을 새로 불러 올 수 있다.

나. Drafting Toolbar -Add New Label – Add New Decal Label -Size (Label 크기), Line width (Label 두께) – OK

- 부품의 Lable의 레이어, 크기, 두께를 조절할 수 있다.

Library 이해 (Drafting Toolbar)

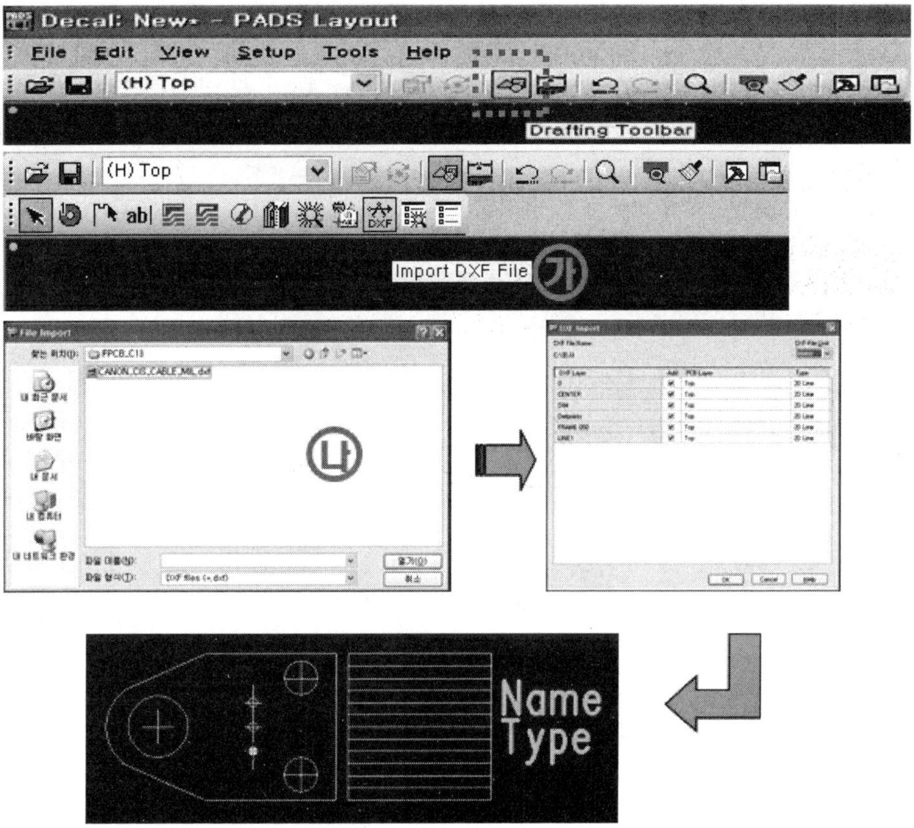

Import DXF File Mode

가. Drafting Toolbar -Import DXF File

- 부품의 외형 기구 도면을 불러올 수 있다.

나. Drafting Toolbar -Import DXF File -File Import (파일형식 .dxf 선택) -열기 DXF Import – OK

- 기구도면은 .dxf 파일 형식으로 열 수 있다.

Library 이해(Drafting Toolbar)

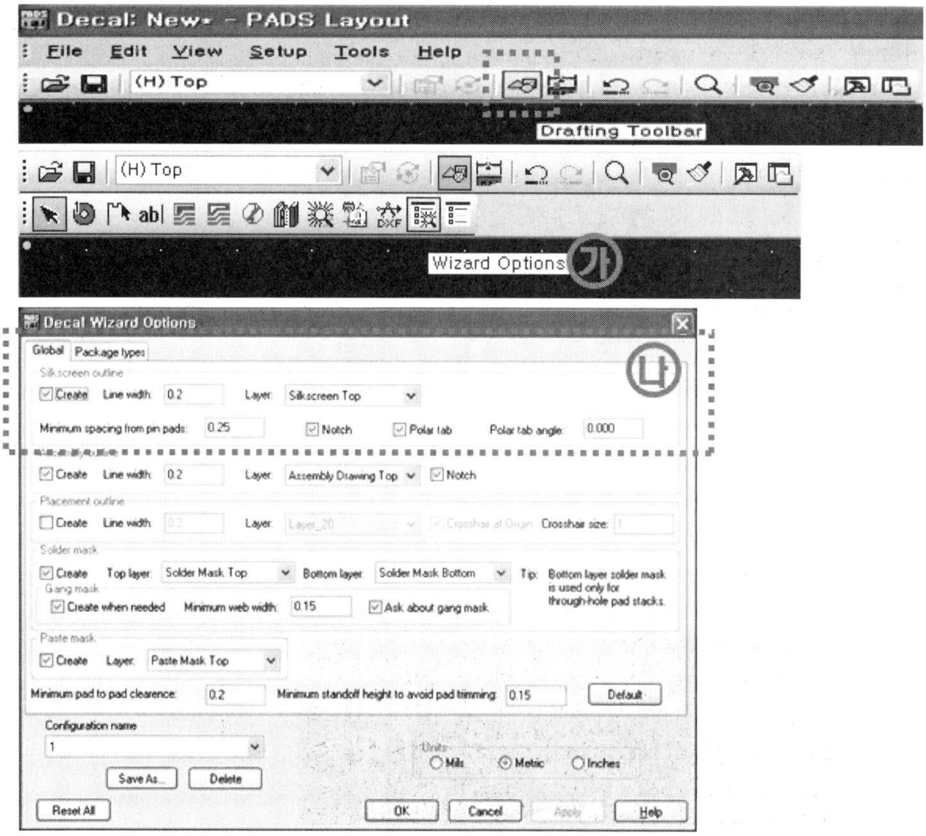

Wizard Options Mode

가. Drafting Toolbar -Wizard Options

- Wizard의 옵션을 변경할 수 있다.

나. Drafting Toolbar -Wizard Options -Decal Wizard Options -Global - Silkscreen Outline - Line Width (부품 외각 두께) -Layer(레이어 선택) - OK

- 부품의 외각의 설정을 변경할 수 있다.

04 / PADS Layout 시작

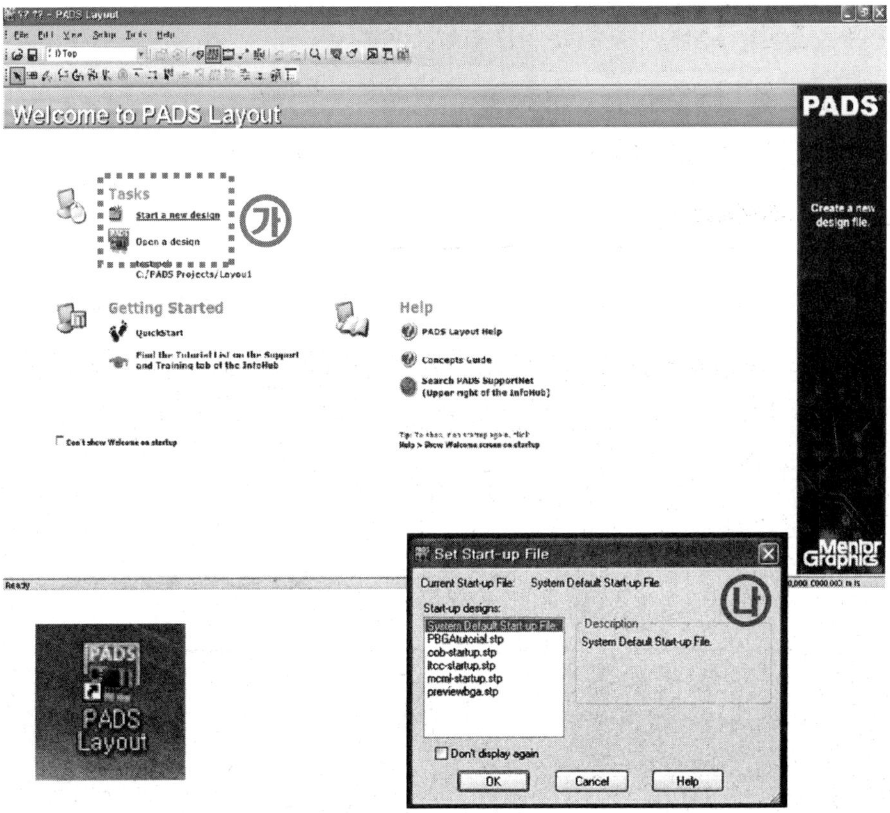

PADS Layout 시작하기

가. PADS Layout Icon – start a new design

나. Set Start-up File – Start-up Design – 선택 – OK

- User가 setting한 기본 창으로 들어간다. (처음일 경우 선택된 Start-up으로 OK)

05 / PCB Decal 작성

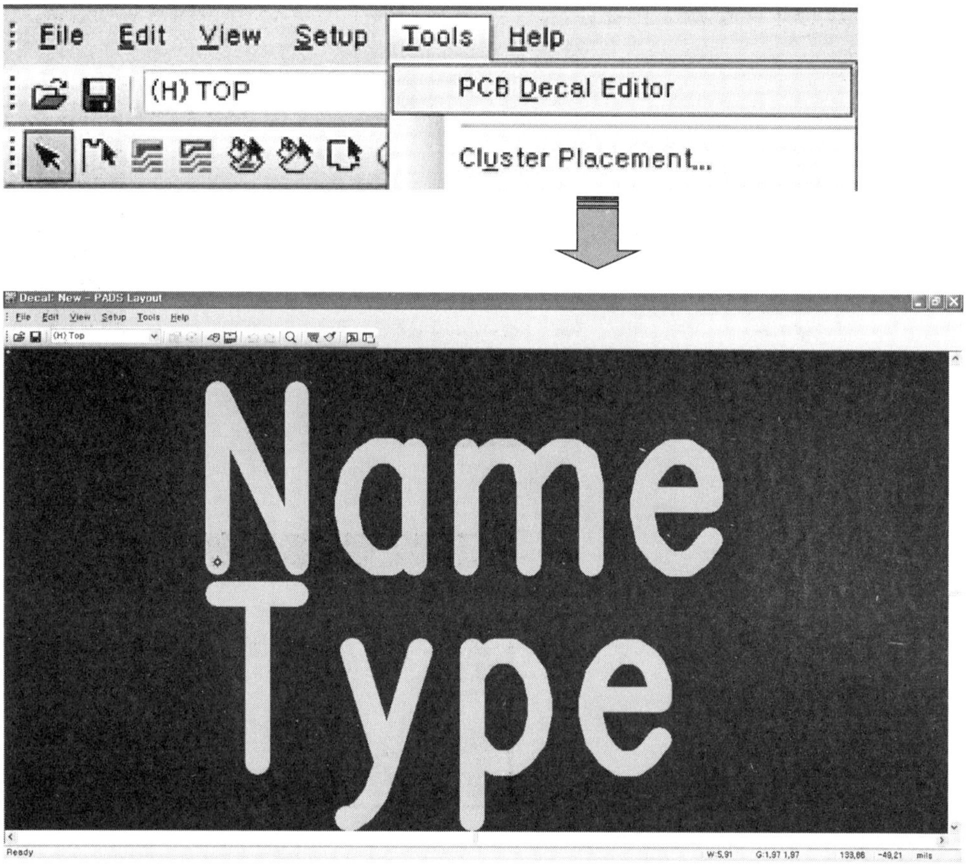

Library로 들어가기

가. Tools - PCB Decal Editor

- 기본 창에서 Library로 들어가 Decal 작성을 한다.
- Library 화면에서는 부품의 Name과 Type이 원점에 표시된다.

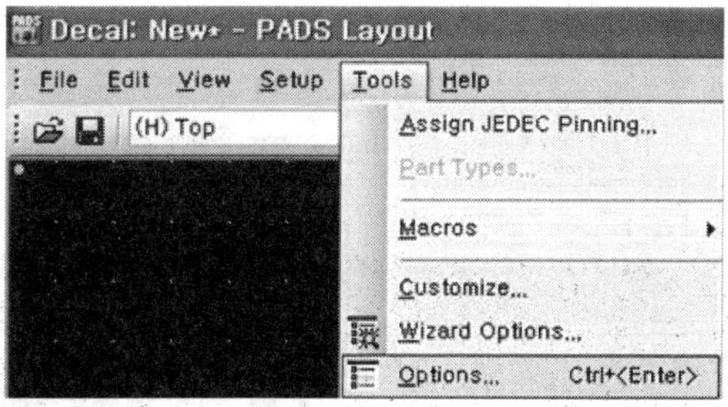

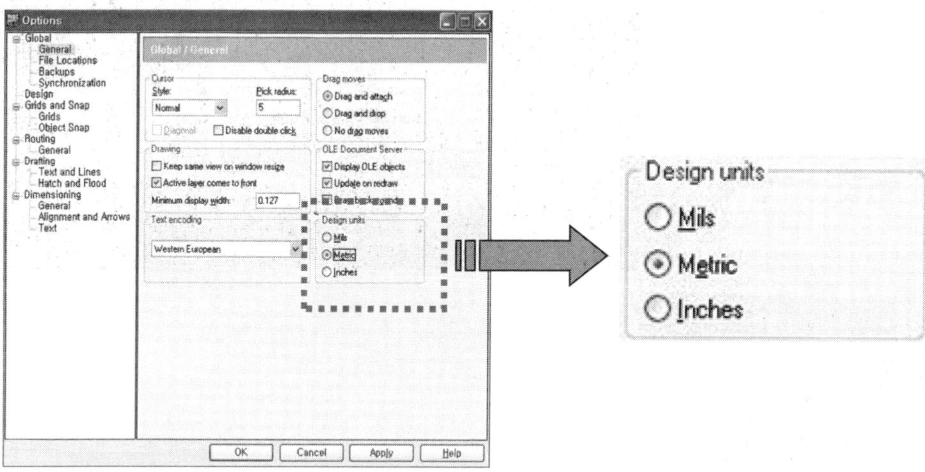

DECAL 작성시 단위변경

가. Tools – PCB Decal Editor 선택

- Main 화면에서 Library 창으로 들어간다.

나. Tools – Option – Global – General – Design units 선택

- Decal 작성시 단위 설정을 한다. (Mils, Metric (mm), Inches)

Terminal 생성하기

가. Drafting Toolbar – Terminal – Add Terminals – OK – s0 0(원점이동) – Enter – Space bar

- 원점에 1번 Terminal을 생성한다.
- Name과 Type을 원점 근처로 이동시킨다.

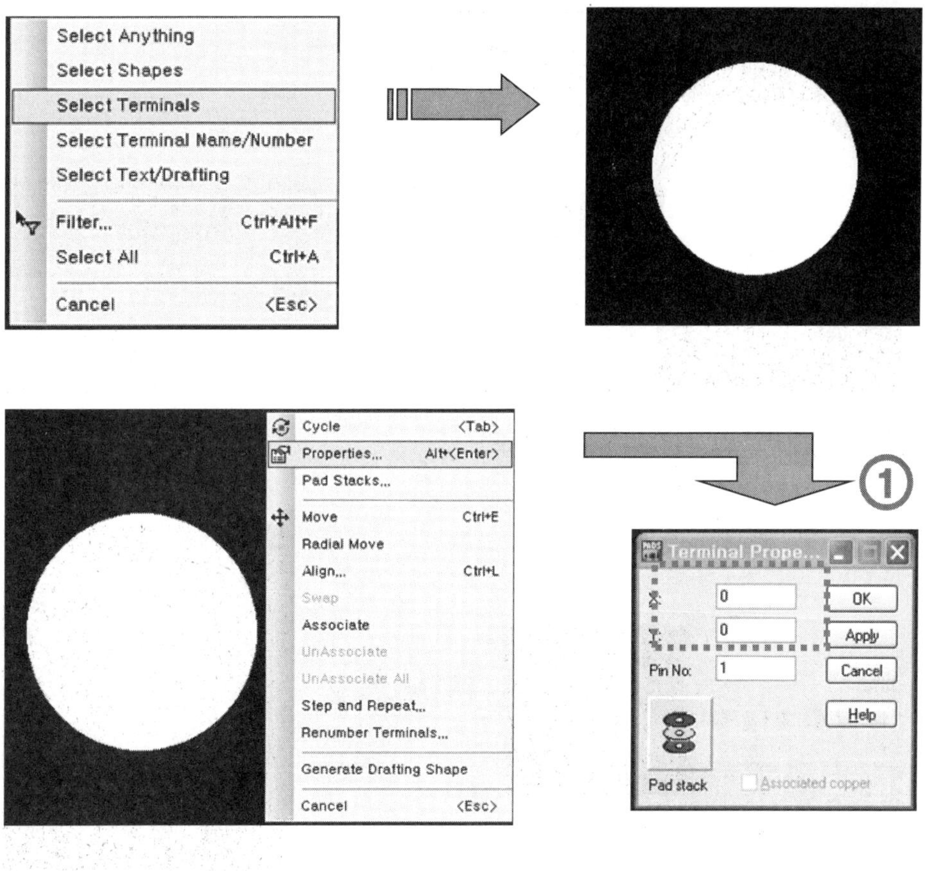

Terminal 위치 확인하기

가. 마우스(우측) – Select Terminals – Terminal 선택 – 마우스(우측) – Properties

- 1번 Terminal이 원점에 위치해 있는지 확인한다. (X/Y 좌표확인)
- 잘못된 위치에 핀이 있을 경우 ①의 그림과 같이 X/Y 좌표를 수정한다.

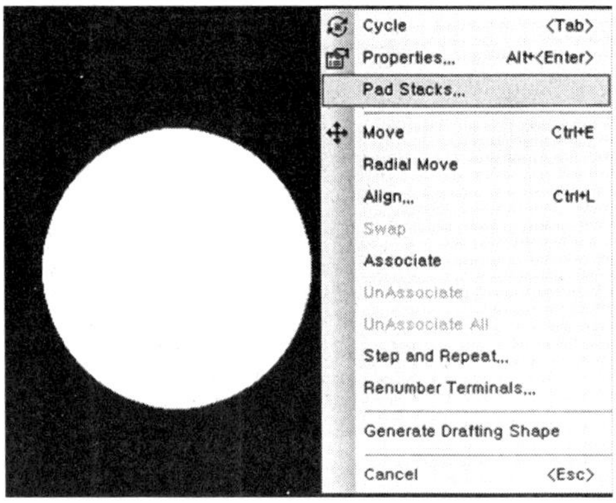

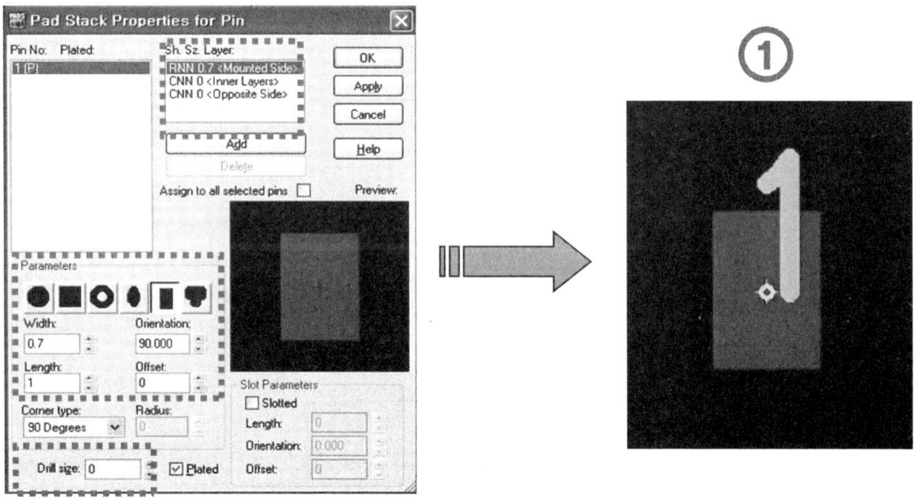

Terminal Size 변경

가. 마우스(우측) – Select Terminals – Terminal 선택 – 마우스(우측) – Pad Stacks

- 1번 Terminal의 Size를 ①과 같이 변경한다.

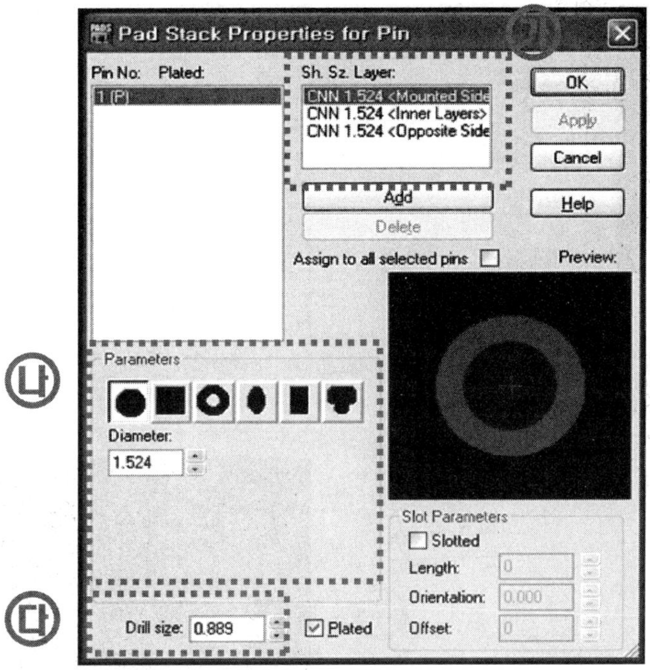

Terminal Size 변경

가. Layer별 Terminal 변경

- DIP Pin의 경우 : Mounted Side, Inner, Opposite Side를 동일하게 변경한다.
- SMD Pin의 경우 : Mounted Side만 변경한다.

나. Parameters – Diameter Size 변경

- Land의 모양 및 사이즈를 변경한다.

다. Drill Size 변경

- DIP Pin의 경우 : 홀 Size 치수를 기입한다.
- SMD Pin의 경우 : 0으로 변경한다.

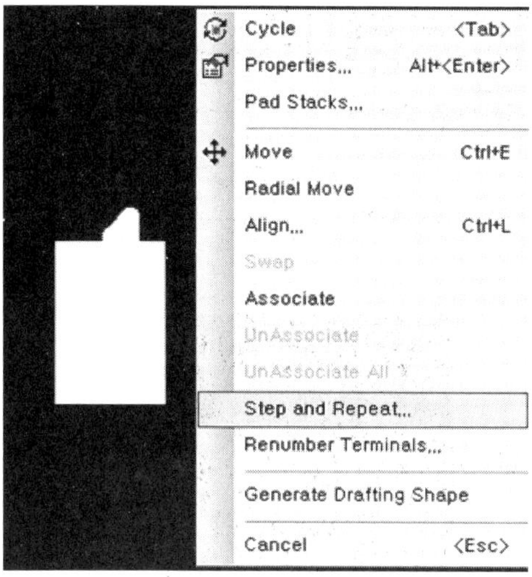

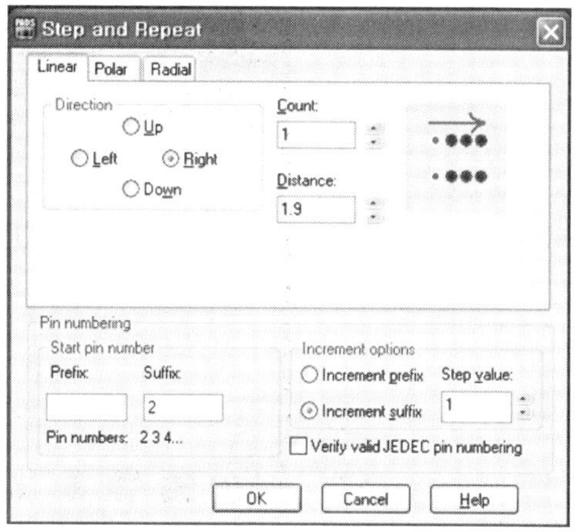

①

Terminal 추가

가. 마우스(우측) – Select Terminals – Terminal 선택 – 마우스(우측) – Step and Repeat – Direction 선택 – Count 선택 – Distance(Pin Pitch) – OK

- ①그림과 같이 1번을 기준으로 2번 Terminal을 추가한다.

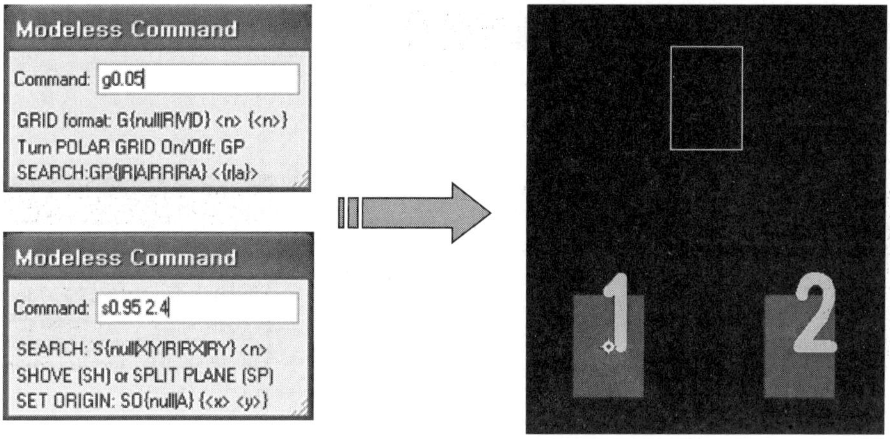

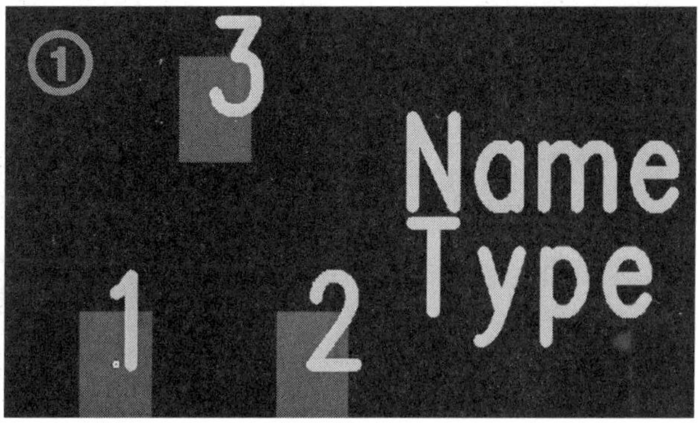

Terminal 추가

가. 마우스(우측) – Select Terminals – Terminal 선택 – 복사 – g0.05 – s0.95 2.4 – Enter – Space bar

- ①그림과 같이 원점을 기준으로 3번 Terminal을 추가한다.

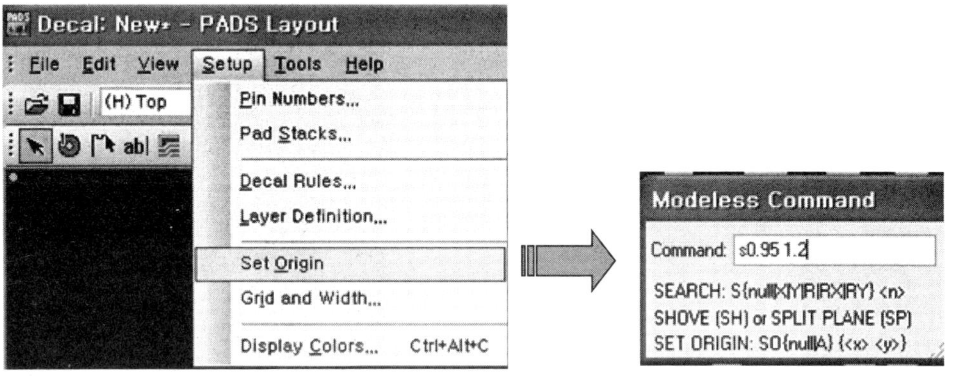

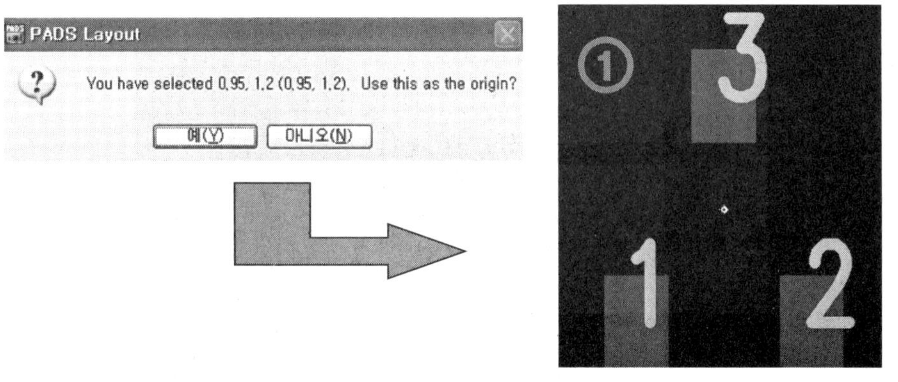

PCB Decal 원점 변경

가. Setup – Set Origin – s0.95 1.2 – Enter – Space bar – Yes

- ①그림과 같이 원점을 변경한다.

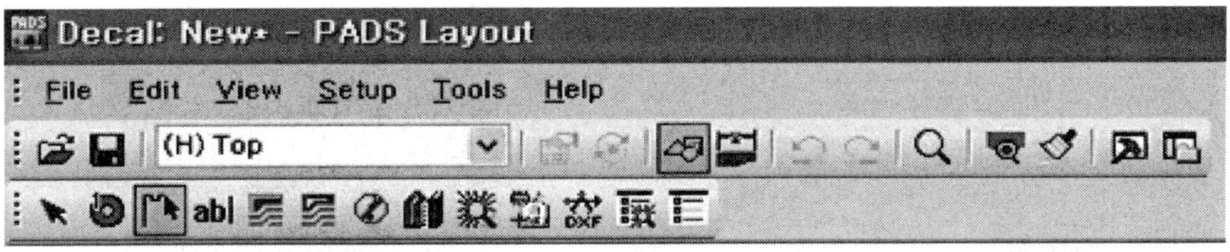

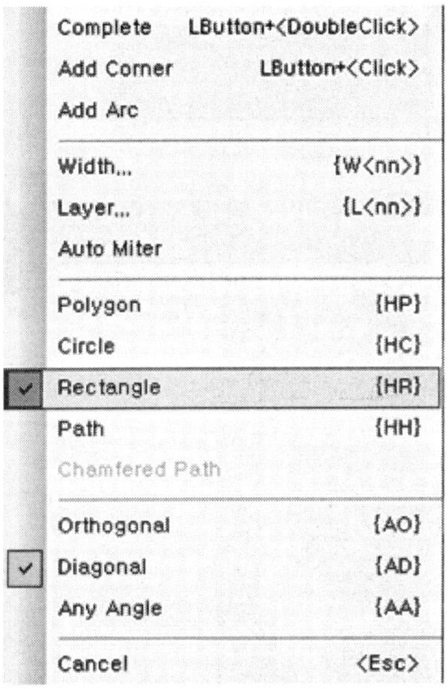

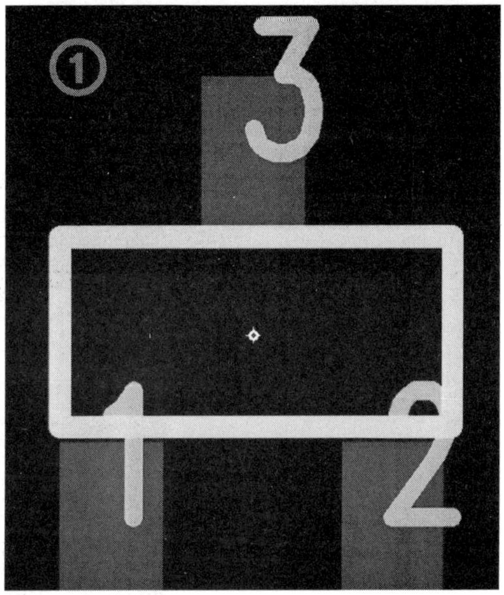

PCB Decal 외형 생성

가. Drafting Toolbar – 2D Line – 마우스(우측) – Rectangle – Box 생성

- ①그림과 같이 Box를 그린다.

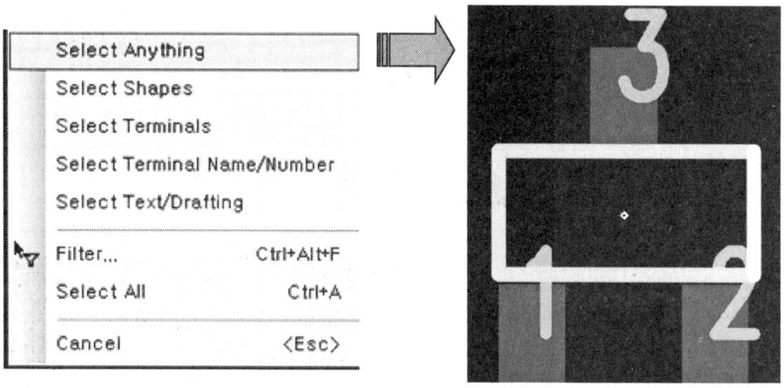

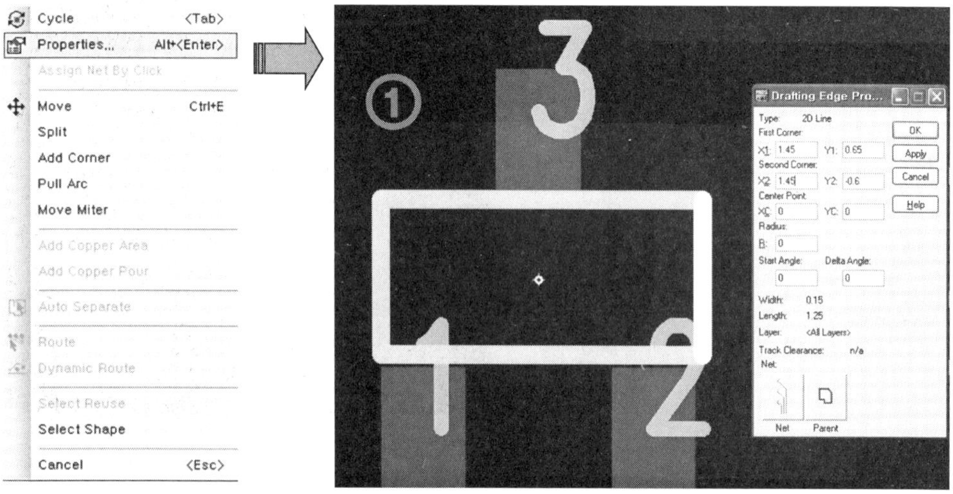

PCB Decal 외형 수정

가. 마우스(우측) – Select Anything – 2D Line 선택 – 마우스(우측) – Properties – Drafting Edge Properties – First Corner/Second Corner 변경

- ①그림과 같이 Box의 Size를 수정한다.

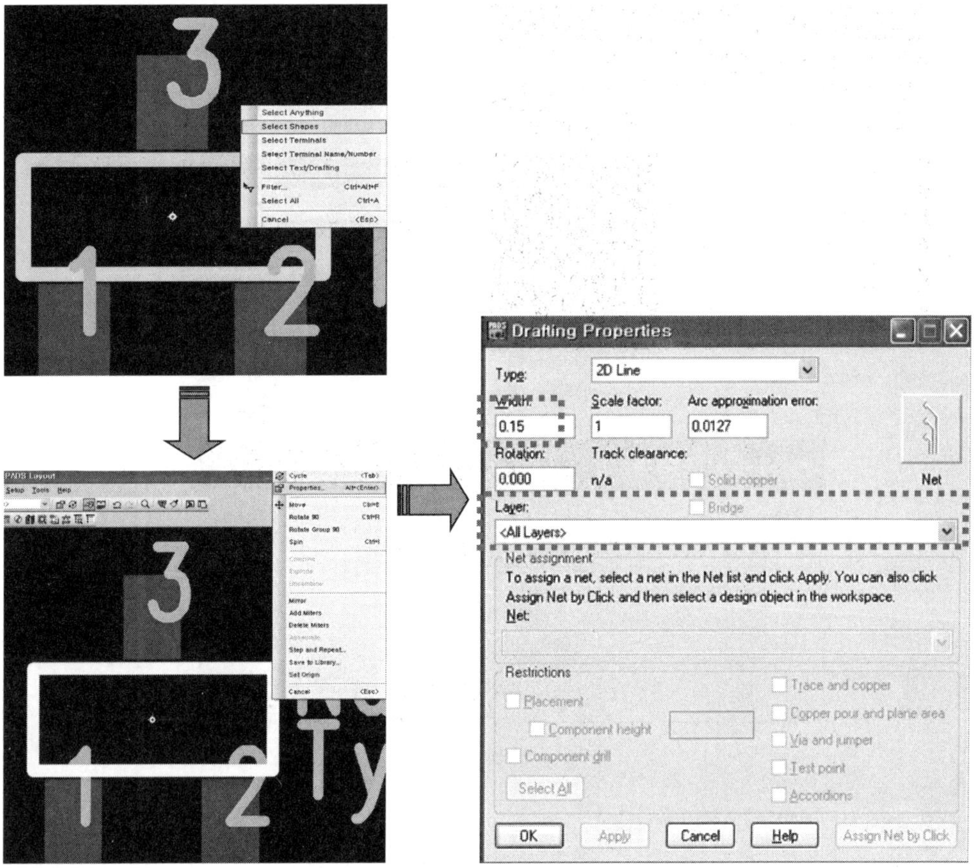

PCB Decal 외형 설정 변경

가. 마우스(우측) – Select Shape – 2D Line 선택 – 마우스(우측) – Properties – Drafting Properties – Width 변경(Line 두께) – Layer 선택(All Layer)

- 부품 외형의 두께 및 Layer를 설정한다.

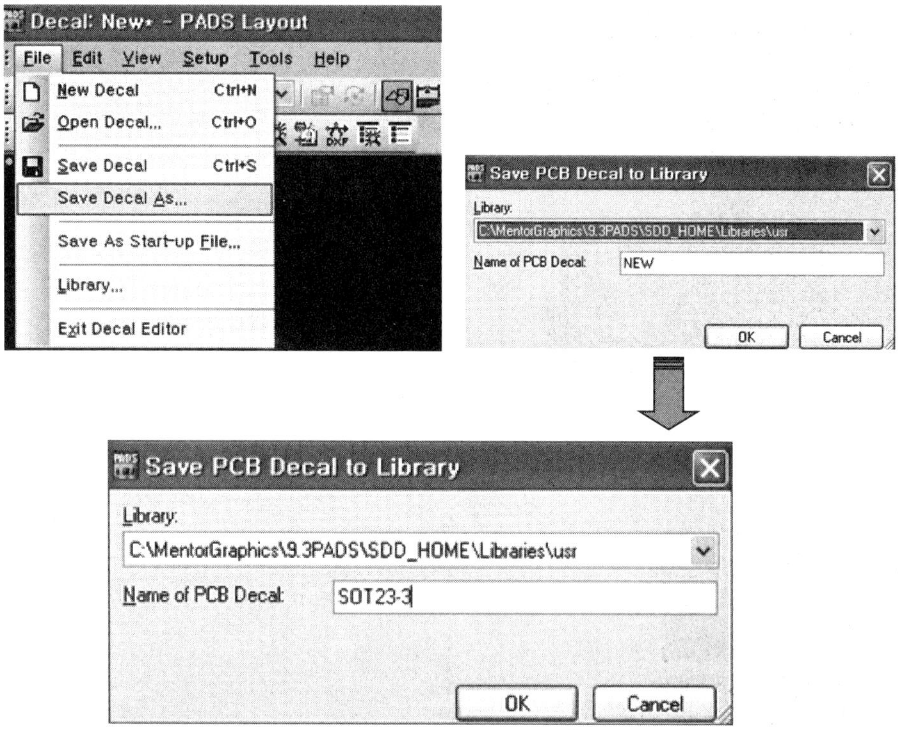

PCB Decal 저장

가. File – Save Decal As... – Save PCB Decal to Library – Decal name (SOT23–3) – OK

- 선택한 Library에 PCB Decal을 저장한다.

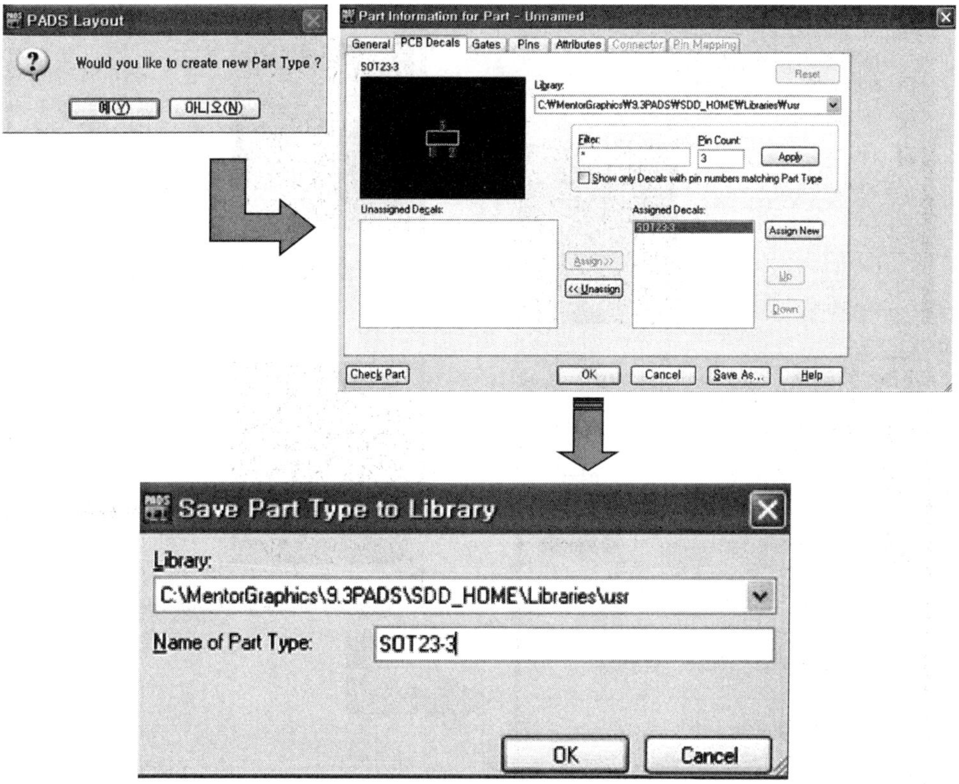

Part Type 저장

가. PCB Decal 저장 – New Part Type – Yes – PCB Decal – Assigned Decals (Decal 확인) – OK – Part Type name (SOT23–3) – OK

- 선택한 Library에 Part Type을 저장한다.

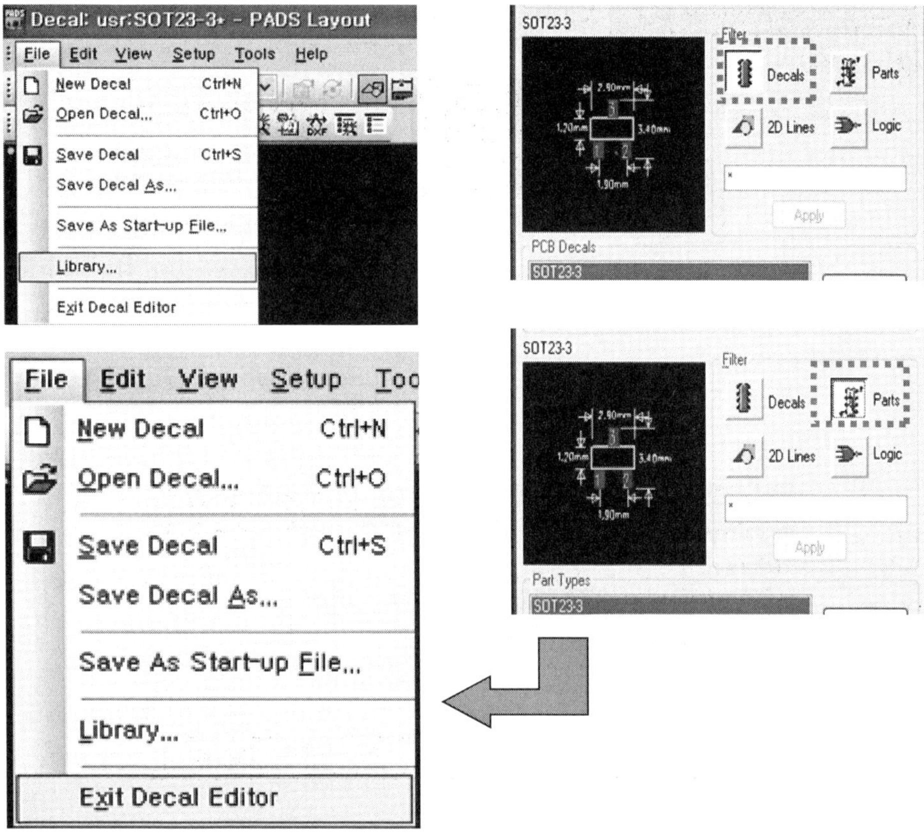

Library 확인

가. File – Library – Library Manager – Library (usr 선택) – Decal/Part Type 확인 – Exit Decal Editor

- Library에 Decal/Part Type이 저장되어 있는지 확인한다.

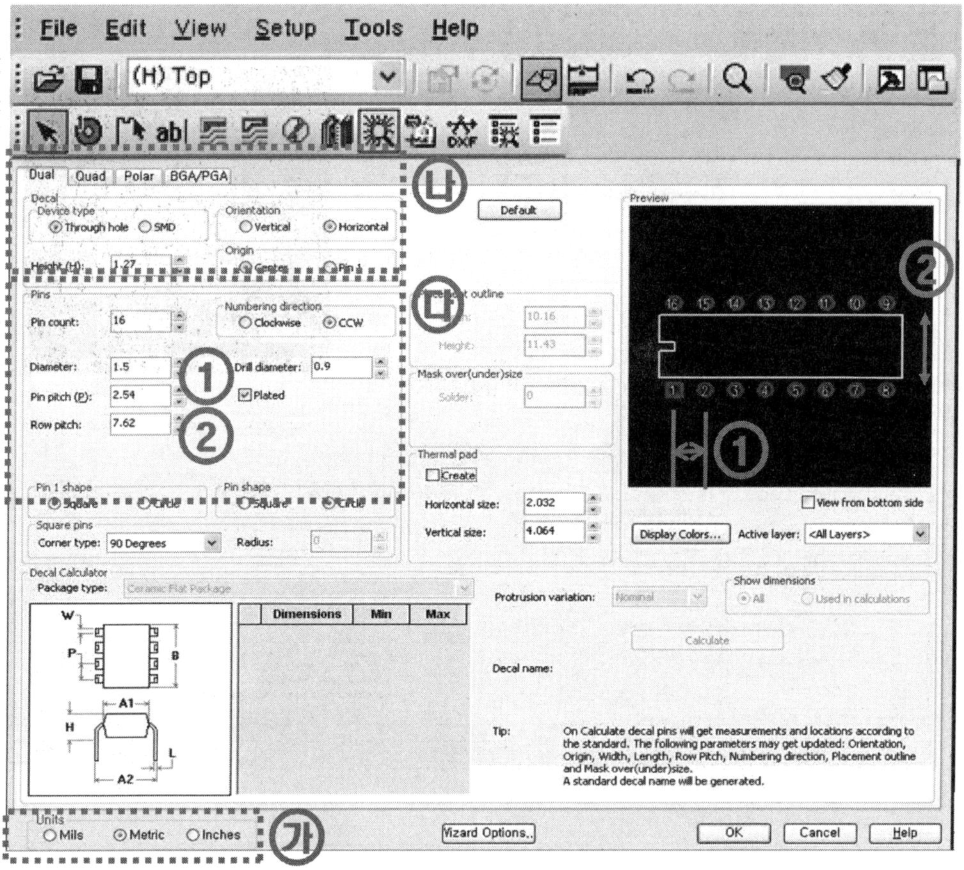

Decal Wizard 설정 (DIP IC)

가. Drafting Toolbar – Wizard – Unit (Metric)

나. Dual 선택 – Device Type (Through hole) – Origin (Center)

- 부품 방향 설정 및 Origin을 설정한다.

다. Pin – Pin Count (16) – Numbering direction (Clockwise=시계 방향, CCW=반 시계 방향) – Diameter/Drill (Land size=1.5/Hole=0.9) – ① Pin Pitch (2.54)

- ② Row Pitch (2개의 열 간격=7.62) - Pin 1 Shape (Square) 선택 – OK

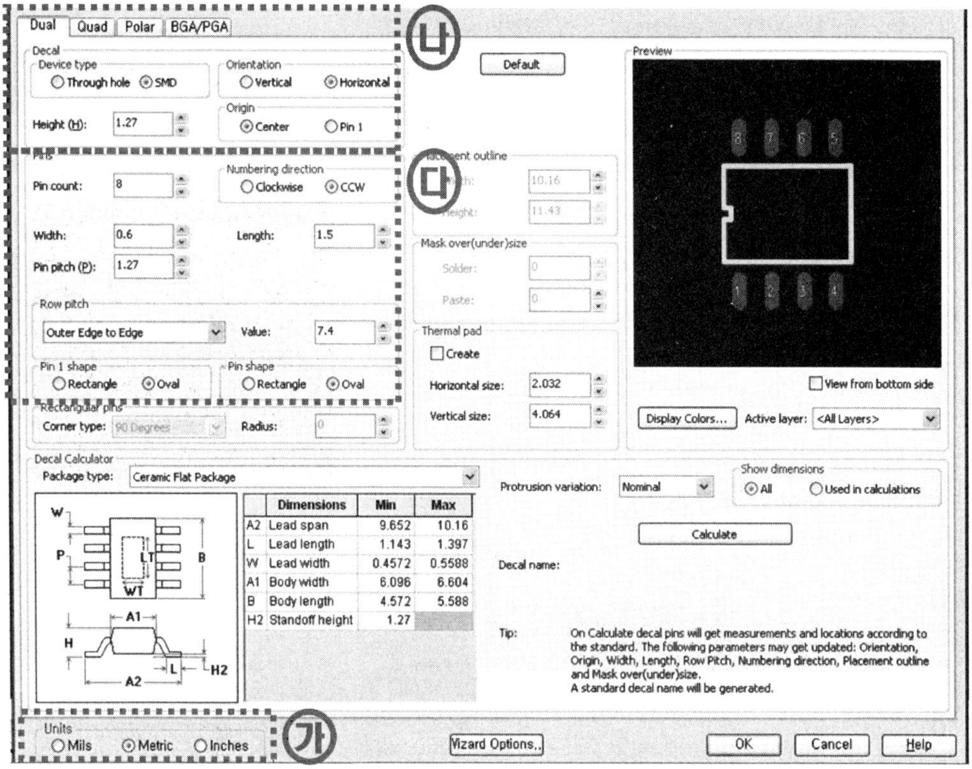

Decal Wizard 설정 (SMD IC)

가. Drafting Toolbar – Wizard – Unit (Metric)

나. Dual 선택 – Device Type (SMD 선택) – Origin (Center)

- 부품 방향설정 및 Origin을 설정한다.

다. Pin – Pin Count (8) – Numbering direction (Clockwise=시계 방향, CCW=반 시계 방향) – Width/Length (Land size=0.6/1.5) – Pin Pitch (1.27) – Row Pitch (2개의 열 간격=Outer Edge7.4) – Pin Shape (Oval) 선택 – OK

- Outer Edge=Pin 바깥쪽 간격, Inner Edge=Pin 안쪽 간격, Center Edge=Center

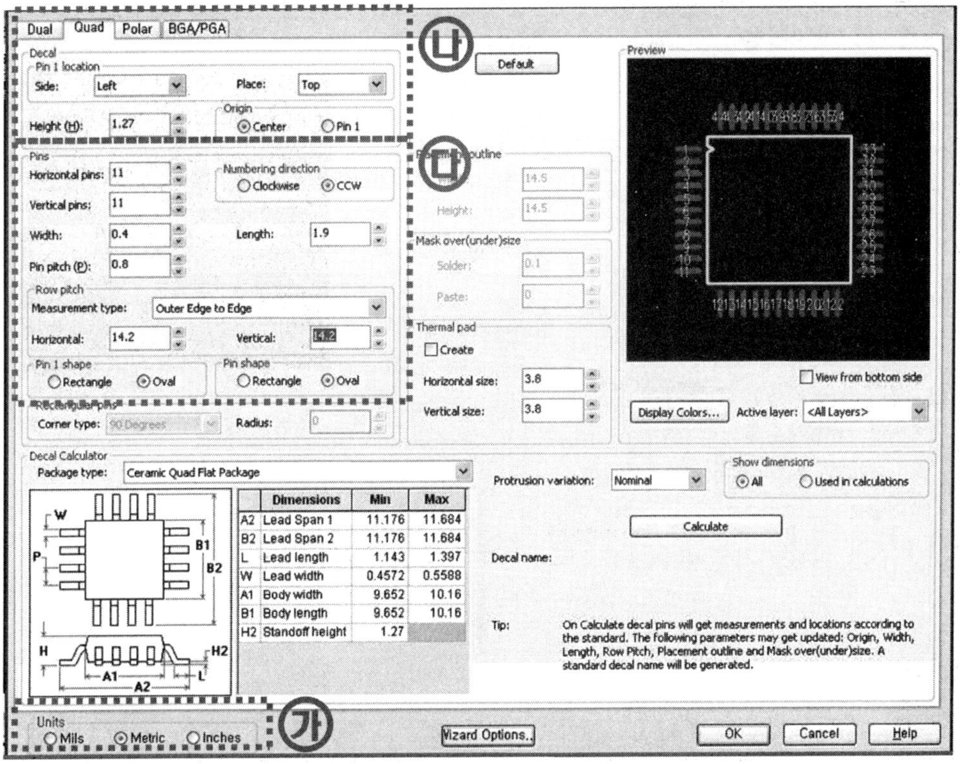

Decal Wizard 설정 (Quad IC)

가. Drafting Toolbar – Wizard – Quad – Unit (Metric)

나. Decal – Pin 1 Location – Side (4열 중 선택=Left)/Place (시작점=Top) – Origin

- 부품 방향설정 및 Origin을 설정한다.

다. Pin – Horizontal Pin (가로 Pin11) – Vertical Pin (세로 Pin11), Numbering direction (Clockwise=시계 방향, CCW=반 시계 방향) – Width/Length (Land size= 0.4/1.9) – Pin Pitch (0.8) – Row Pitch (2개의 열 간격=Outer Edge) – Horizontal (14.2) Vertical (14.2) – Pin Shape (Oval) 선택 – OK

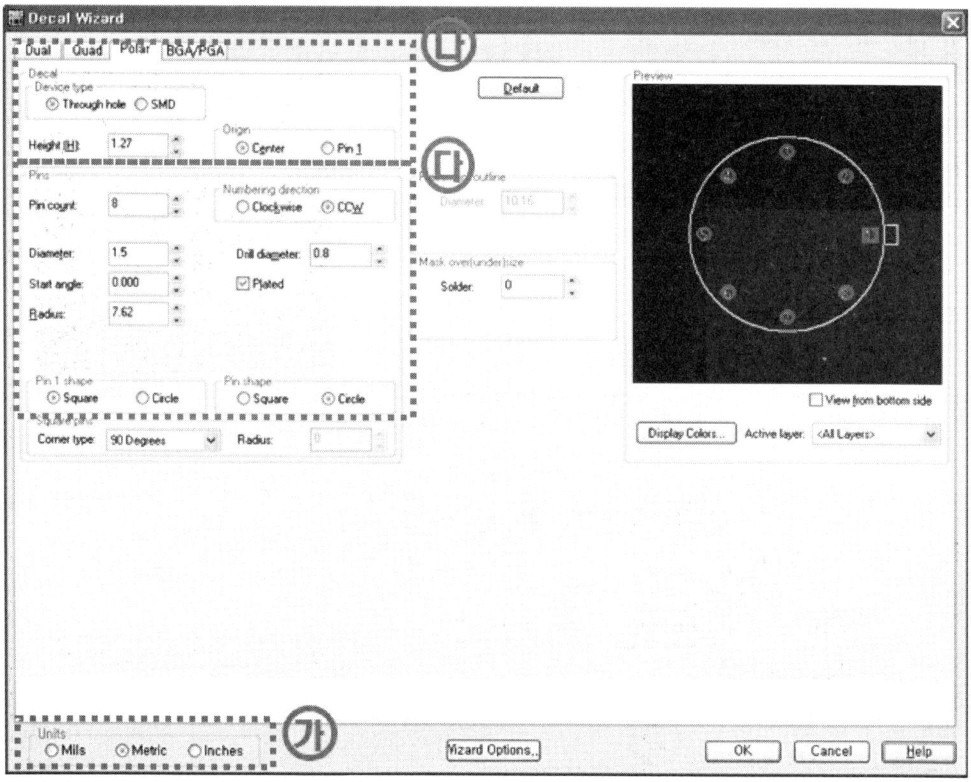

Decal Wizard 설정 (Polar)

가. Drafting Toolbar -Wizard -Polar - Unit (Metric)

나. Device Type (Through hole or SMD) - Origin (Center)

- 부품의 Type 및 Origin을 설정한다.

다. Pin -Pin Count (8) -Numbering direction (Clockwise=시계 방향, CCW= 반시계 방향) -Diameter / Drill (Land size=1.5/Hole=0.8) -①Start Angle(90) -②Radius(반지름 거리=7.62) -Pin 1 Shape (Square), Pin Shape (Circle) -선택 - OK

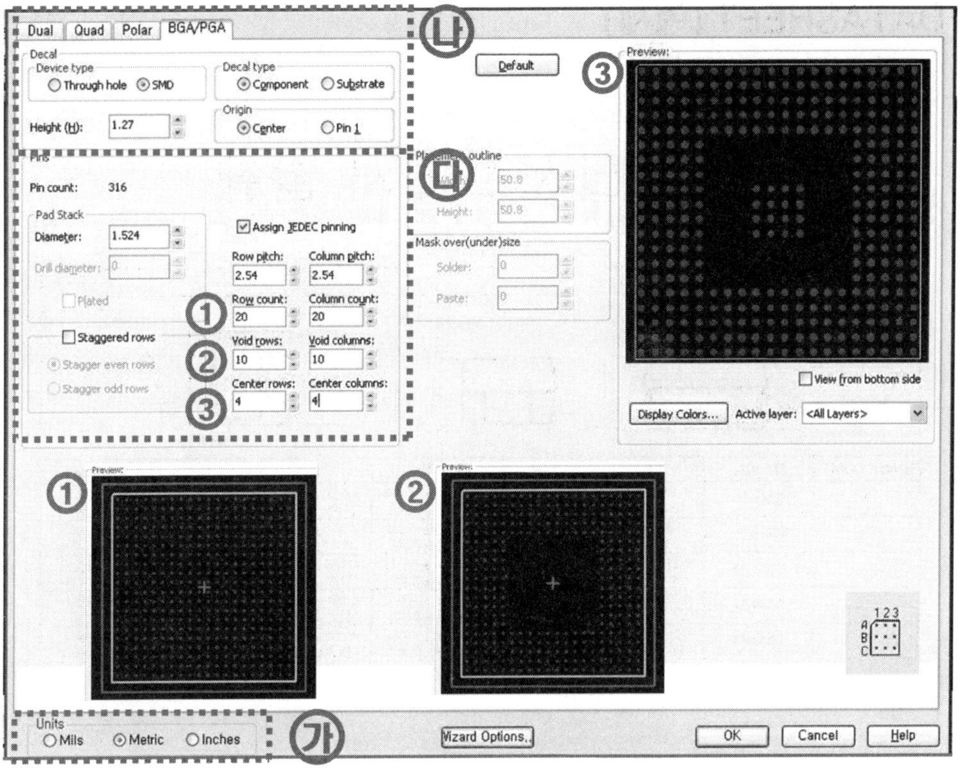

Decal Wizard 설정 (BGA/PGA IC)

가. Drafting Toolbar – Wizard – BGA/PGA – Unit (Metric)

나. Decal – Device Type (Through hole=DIP/SMD=Ball Type) – Origin (Center)

- 부품 Pin Type 설정 및 Origin을 설정한다.

다. Pad Stack – Diameter (Land size=1.524) – Pin Pitch (Row=세로/Column=가로) – ① Count (Pin 수=20/20) – ② Void (삭제 Pin수=10/10) – ③ Center (Pin 수=4/4) – OK

DATASHEET (예제)

Vishay 데이타시트에서 발췌

DIMENSIONS in inches (millimeters)

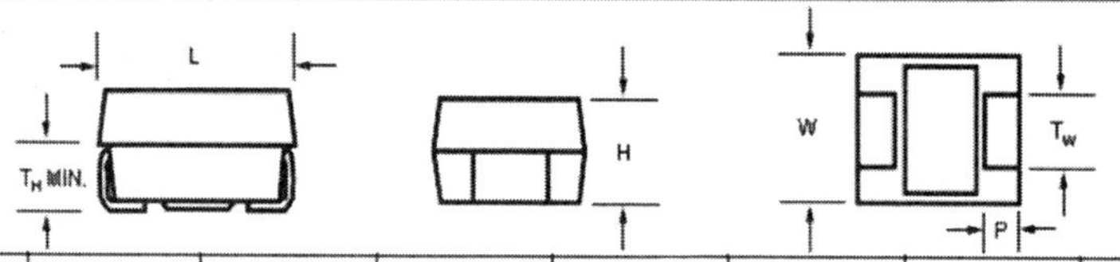

CASE CODE	EIA SIZE	L	W	H	P	T_W	T_H (MIN.)
A	3216-18	0.126 ± 0.008 [3.2 ± 0.20]	0.063 ± 0.008 [1.6 ± 0.20]	0.063 ± 0.008 [1.6 ± 0.20]	0.031 ± 0.012 [0.80 ± 0.30]	0.047 ± 0.004 [1.2 ± 0.10]	0.028 [0.70]
B	3528-21	0.138 ± 0.008 [3.5 ± 0.20]	0.110 ± 0.008 [2.8 ± 0.20]	0.075 ± 0.008 [1.9 ± 0.20]	0.031 ± 0.012 [0.80 ± 0.30]	0.087 ± 0.004 [2.2 ± 0.10]	0.028 [0.70]
C	6032-28	0.236 ± 0.012 [6.0 ± 0.30]	0.126 ± 0.012 [3.2 ± 0.30]	0.098 ± 0.012 [2.5 ± 0.30]	0.051 ± 0.012 [1.3 ± 0.30]	0.087 ± 0.004 [2.2 ± 0.10]	0.039 [1.0]
D	7343-31	0.287 ± 0.012 [7.3 ± 0.30]	0.170 ± 0.012 [4.3 ± 0.30]	0.110 ± 0.012 [2.8 ± 0.30]	0.051 ± 0.012 [1.3 ± 0.30]	0.095 ± 0.004 [2.4 ± 0.10]	0.039 [1.0]
E	7343-43	0.287 ± 0.012 [7.3 ± 0.30]	0.170 ± 0.012 [4.3 ± 0.30	0.158 ± 0.012 [4.0 ± 0.30]	0.051 ± 0.012 [1.3 ± 0.30]	0.095 ± 0.004 [2.4 ± 0.10]	0.039 [1.0]

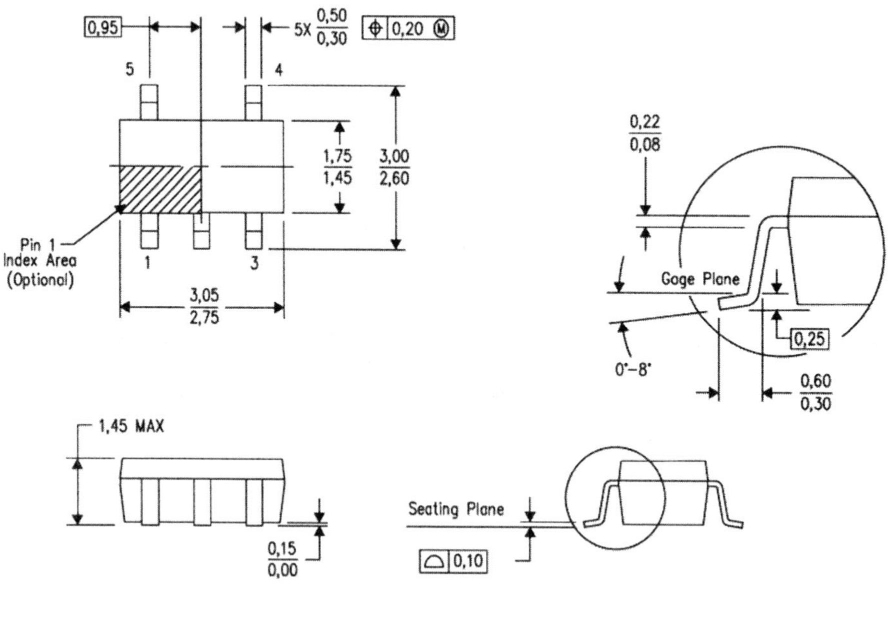

• QFP32

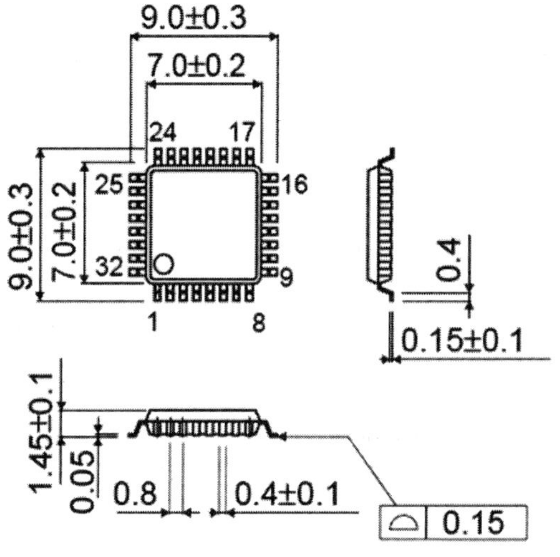
9.0±0.3
7.0±0.2
24
17
25
16
32
9
1
8
9.0±0.3
7.0±0.2
0.4
0.15±0.1
1.45±0.1
0.05
0.8
0.4±0.1
0.15

• DIP16

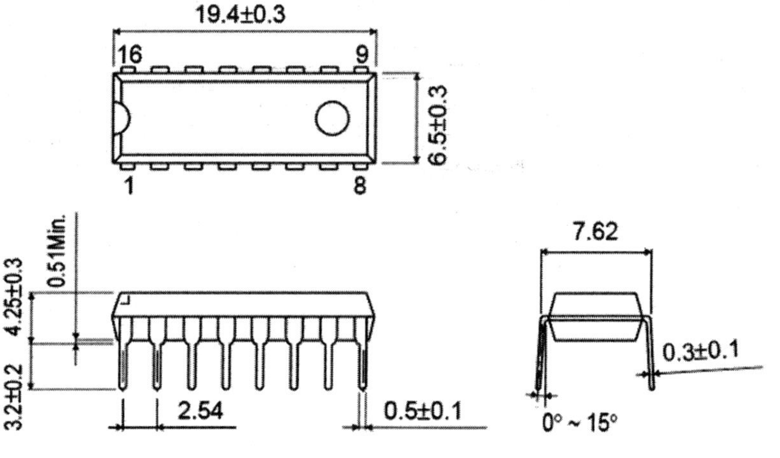
19.4±0.3
16
9
1
8
6.5±0.3
0.51Min.
4.25±0.3
3.2±0.2
2.54
0.5±0.1
7.62
0.3±0.1
0° ~ 15°

Package information - SOT23-6

Surface mounted, 6 pin package

Package outline

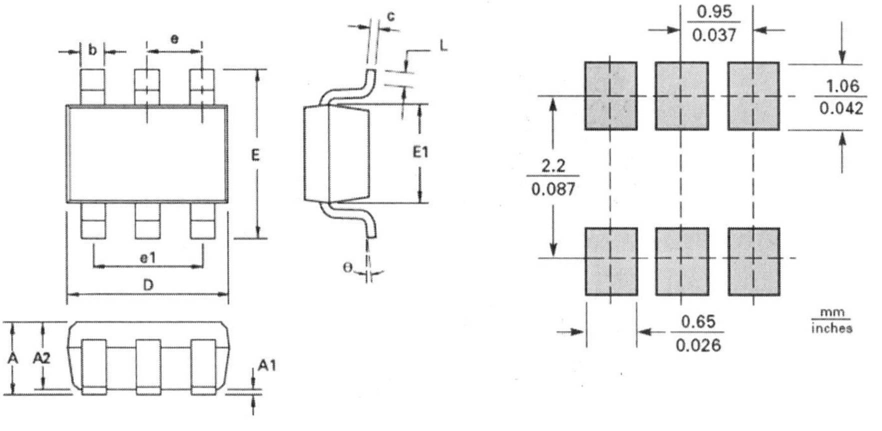

DIM	Millimeters		Inches	
	Min.	Max.	Min.	Max.
A	0.90	1.45	0.0354	0.0570
A1	0.00	0.15	0.00	0.0059
A2	0.90	1.30	0.0354	0.0511
b	0.35	0.50	0.0078	0.0196
C	0.09	0.26	0.0035	0.0102
D	2.70	3.10	0.1062	0.1220
E	2.20	3.20	0.0866	0.1181
E1	1.30	1.80	0.0511	0.0708
L	0.10	0.60	0.0039	0.0236
e	0.95 REF		0.0374 REF	
e1	1.90 REF		0.0748 REF	
L	0°	30°	0°	30°

Note: Controlling dimensions are in millimeters. Approximate dimensions are provided in inches

OR-CAD (PCB Footprint ①)

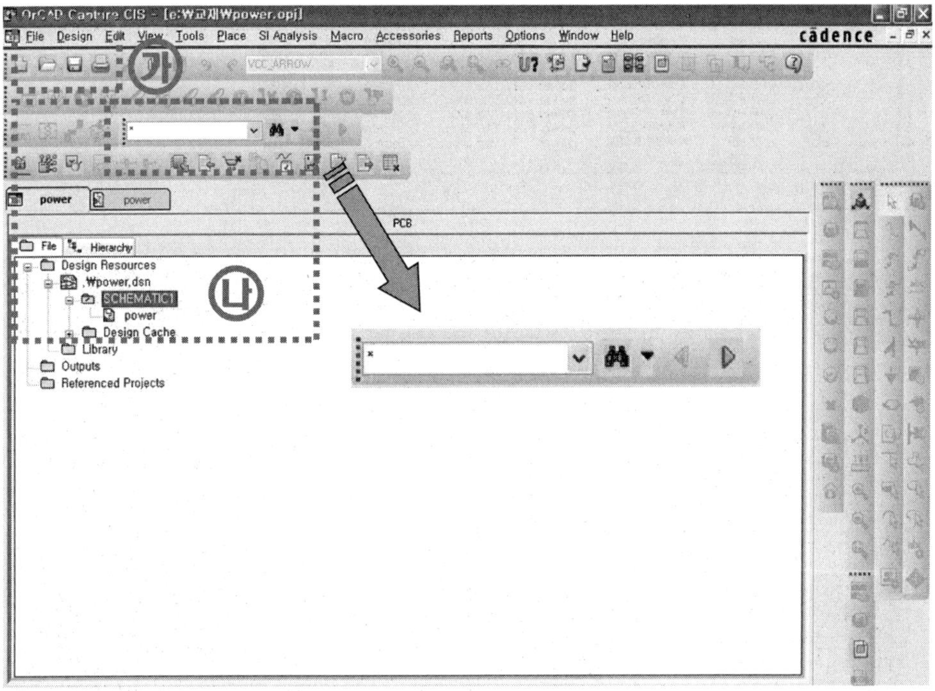

OR-CAD (PCB Footprint 입력하기)

가. OR CAD – File – Open – Design – File 선택

나. File – .dsn(회로도) Double click – SCHEMATIC 선택 – CTRL+F – * – Enter

- 회로에 있는 전체 부품 심볼을 한번에 선택할 수 있다.
- Find 기능은 특정 부품이나 net과 같이 개별로 선택하여 찾기가 가능하다.
- Find 기능을 이용하여 중복된 Ref가 있는 경우 Error를 수정할 수 있다.
- Find 기능을 이용하여 Page별 회로의 연결을 쉽게 찾을 수 있다.

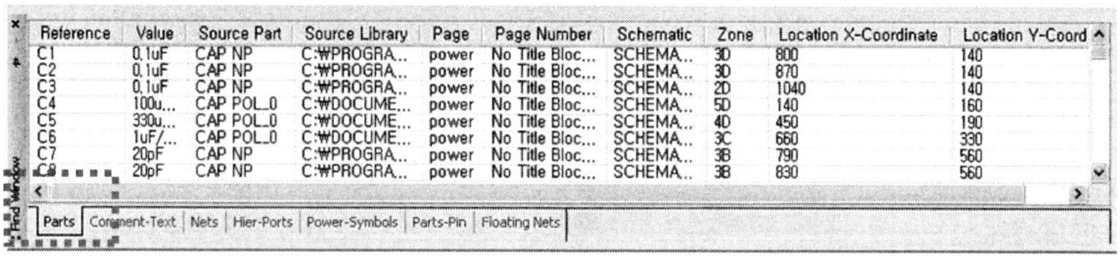

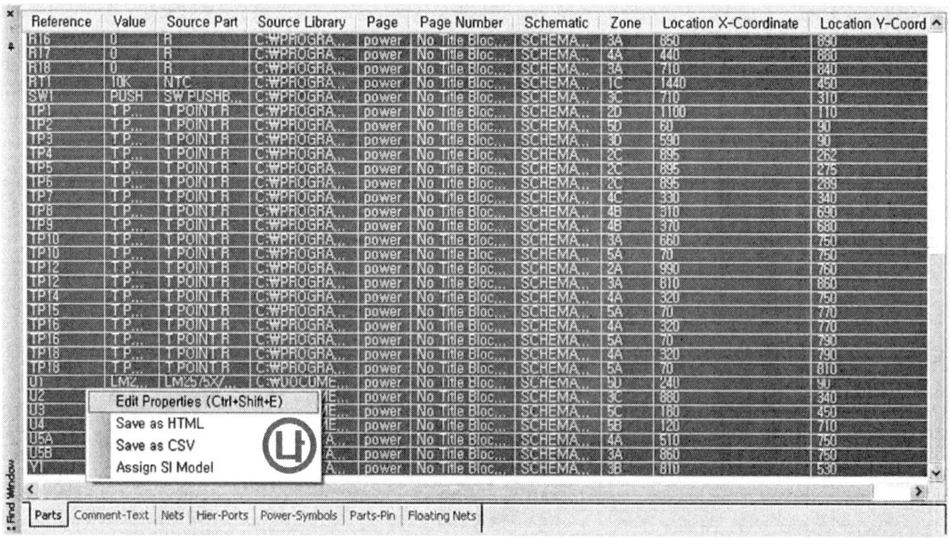

OR-CAD (PCB Footprint 입력하기)

가. CTRL+F – * – Enter – Parts 선택

- 하단 Page 중 Parts를 선택한다.

나. 전체 Reference 선택 – 마우스(우측) – Edit Properties 선택

- 선택된 심볼의 속성 및 Footprint를 변경할 수 있다.

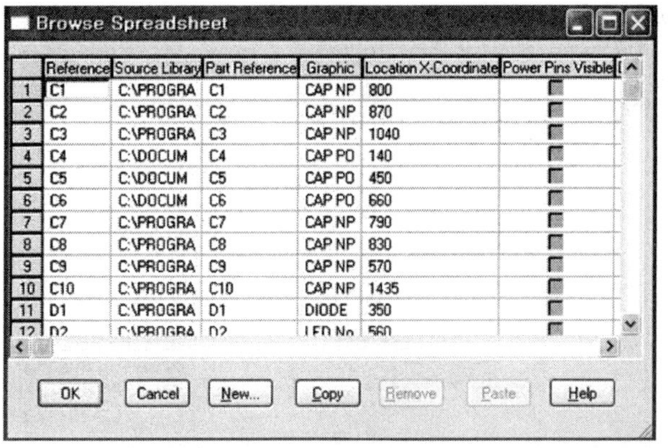

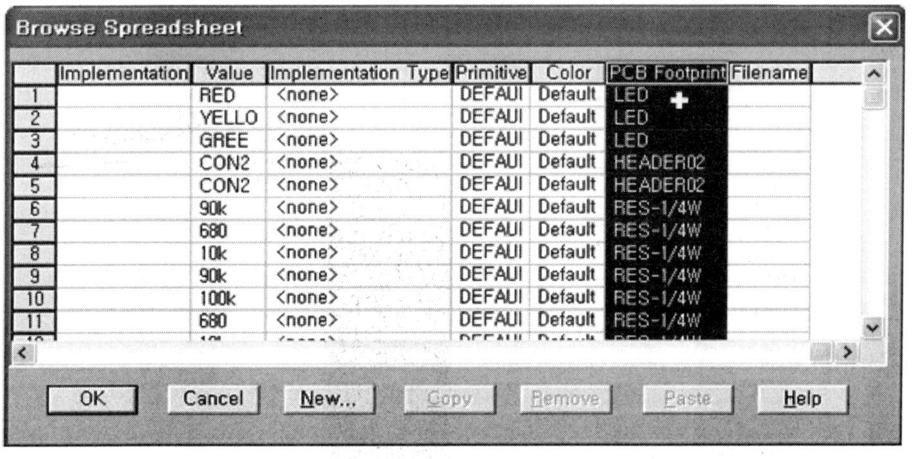

OR-CAD (PCB Footprint 입력하기)

가. Browse Spreadsheet – 우측 끝으로 이동 – PCB Footprint

- 오른쪽으로 이동하면 PCB Footprint를 등록할 수 있다.

나. PCB Footprint

- Layout에 저장된 Part Type으로 등록해야 한다. (주의 : 특수문자, 띄어쓰기 구분)

OR-CAD (PCB Footprint ②)

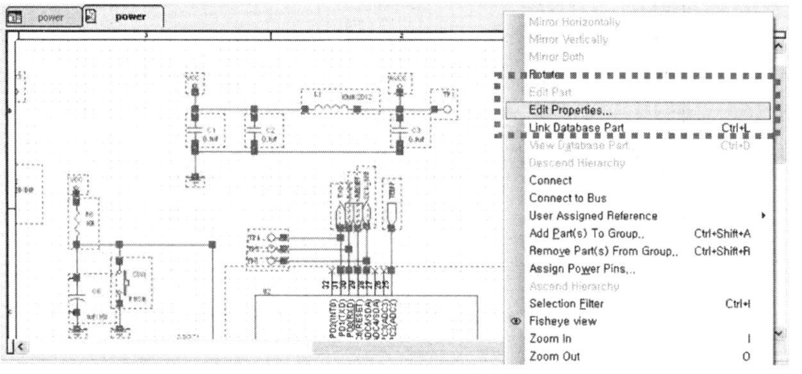

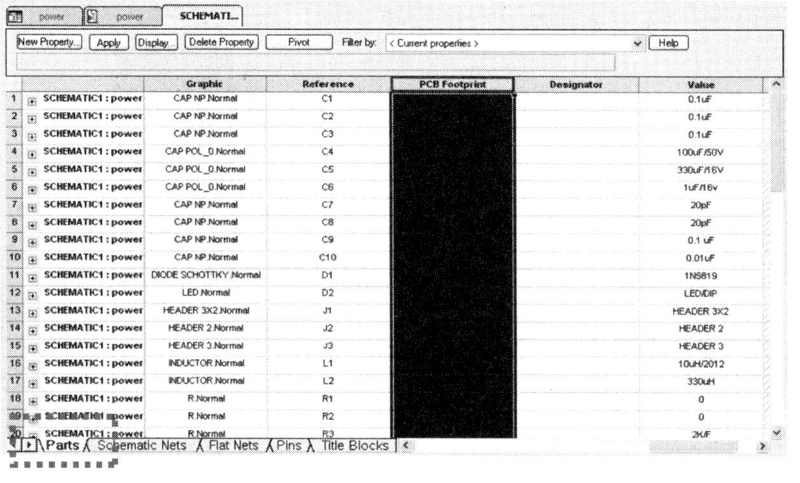

OR-CAD (PCB Footprint 입력하기-Page별)

가. 선택된 Page Open – 전체선택 - 마우스(우측) – Edit Properties – Parts 선택

- 오른쪽으로 이동하면 PCB Footprint를 등록할 수 있다.
- 중복된 Reference가 있는지 확인할 수 있다.
- 같은 Footprint 등록 시 여러 칸을 선택하여 한꺼번에 수정할 수 있다.

OR-CAD(Net list 생성)

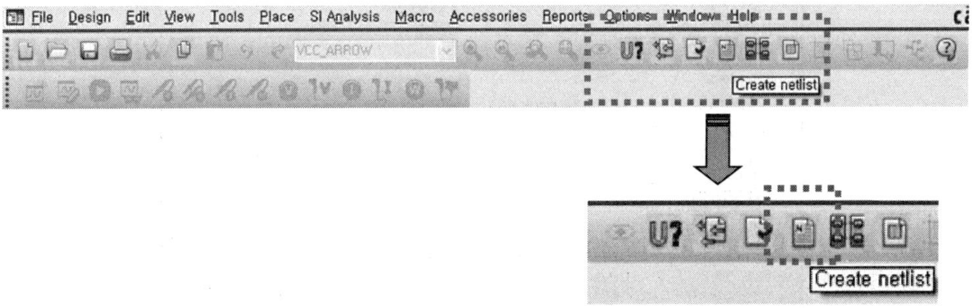

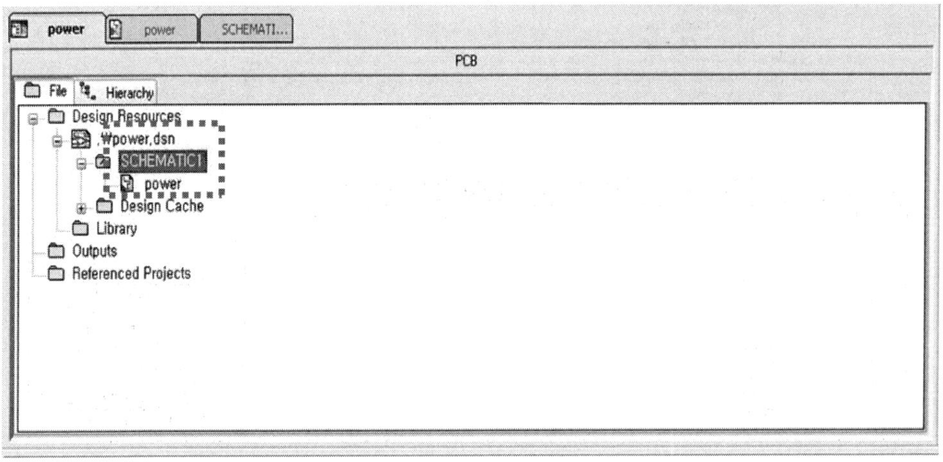

OR-CAD (Net list 생성하기)

가. SCHEMATIC 선택 – Create net list 선택

- Net list : 부품의 Footprint 정보, Net 정보를 포함한 DATA
- 주의 : 회로에 Ref 중복일 경우 Net list가 생성되지 않는다.

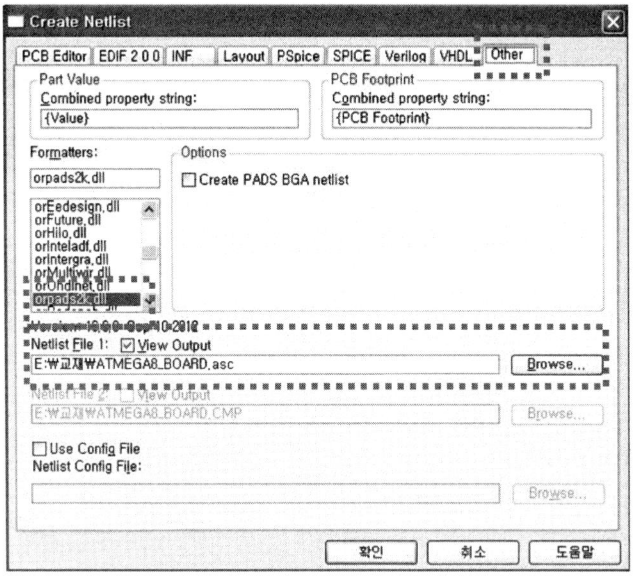

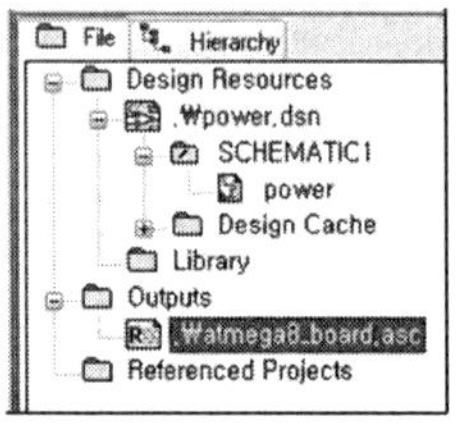

OR-CAD (Net list 생성하기)

가. Create net list - Other 선택 - Formatters (orpads2k.dll) – View Output (확인 시.asc File 열림) – 확인 – XXX.asc

- Output 폴더에서 생성된 XXX.asc를 확인한다.

OR-CAD (BOM 생성)

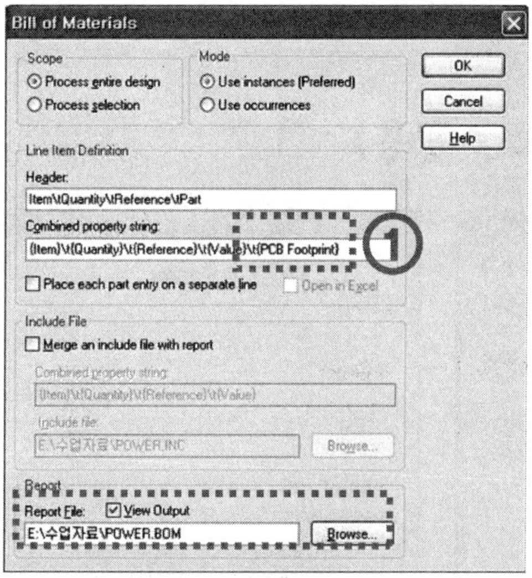

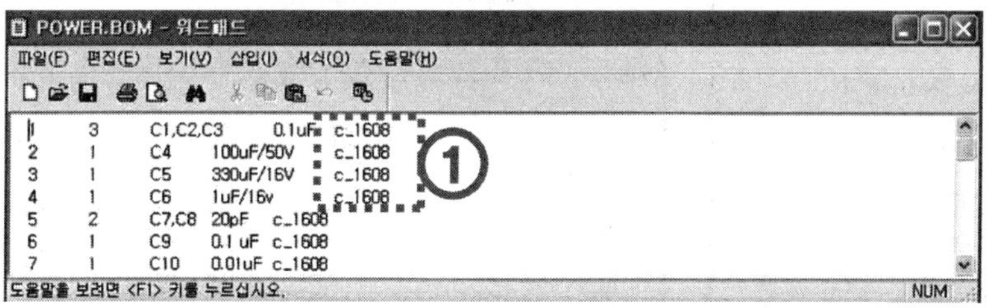

OR-CAD (Net list 생성하기)

가. Bill of Materials – Combined Property string (₩t {PCB Footprint} 입력) – Report (저장경로 및 이름설정) – View Output (확인 시 XXX.BOM File 열림) – OK

- ①의 그림과 같이 Footprint가 추가된 XXX.BOM를 확인한다.

06 / Layout 활용

Layout (Import 활용)

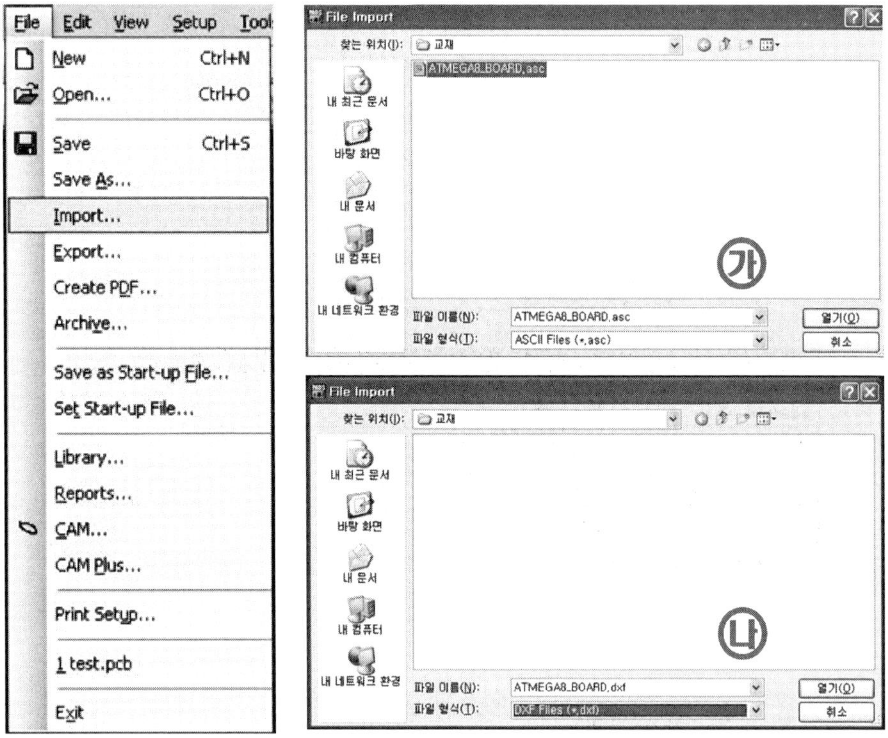

Layout (Import 활용하기)

가. File – Import – File Import (파일형식) – ASCII File

- Net list를 불러 올 수 있다.

나. File – Import – File Import (파일형식) – DXF File

- DXF(기구) File을 불러 올 수 있다.

Layout (Export 활용)

가

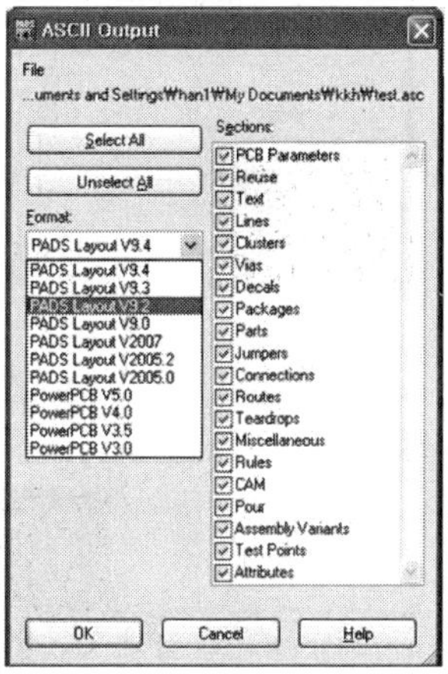

나

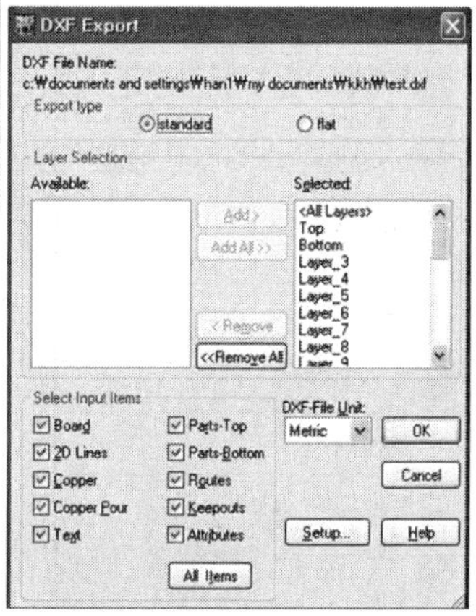

Layout (Export 활용하기)

가. File – Export – File Import (파일형식) – ASCII File – ASCII Output – Select All – Format Down – OK

- Layout의 Version을 다운시켜 낮은 Version에서 작업이 가능하다.

나. File – Export – File Import (파일형식) – DXF File

- DXF(기구) File로 저장할 수 있다.

Layout (Find 활용)

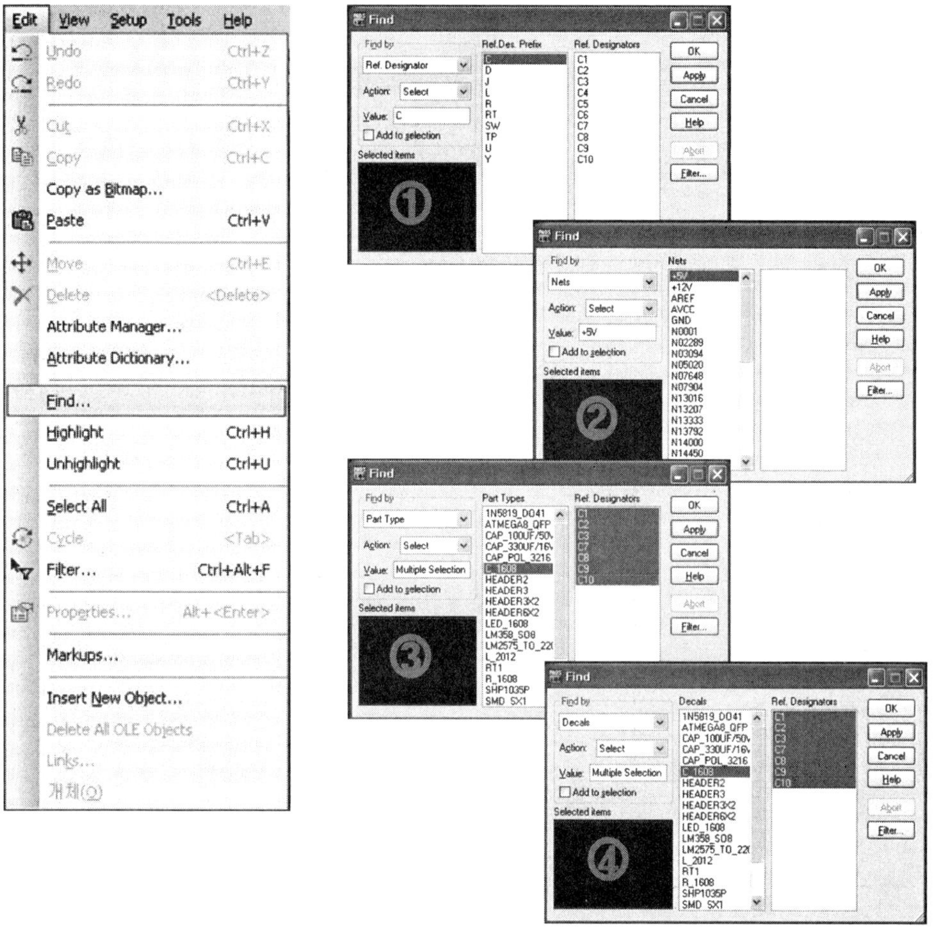

Layout (Find 활용하기)

가. Edit – Find – ① Ref. Designator (Ref별) / ② Net / ③ Part Type / ④ Decal – OK

- 각 파트 별로 지정하여 특정 Item을 선택할 수 있다.

Layout (Filter 활용)

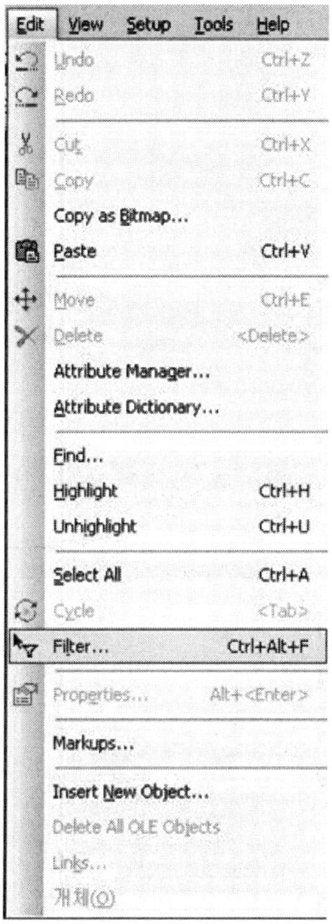

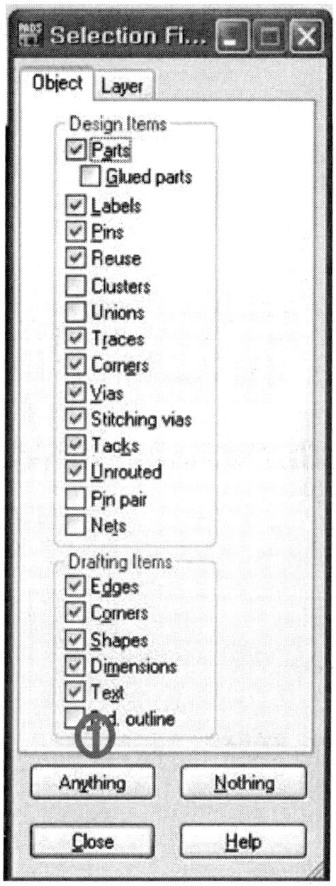

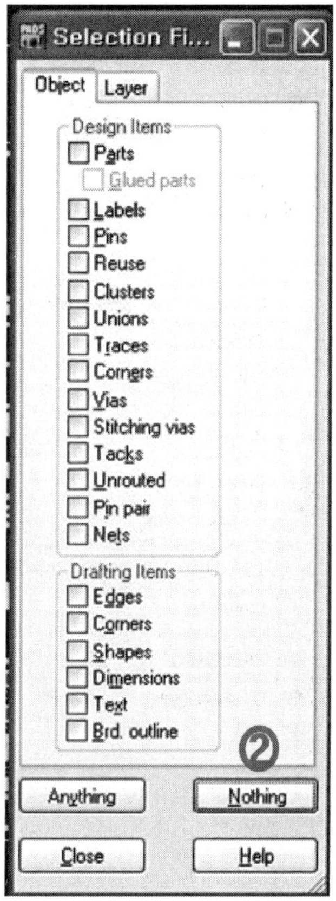

Layout (Filter 활용하기)

가. Edit – Filter – ① Anything / ② Nothing

- User가 원하는 특정 Item을 선택할 수 있다.

Layout (Nets 활용)

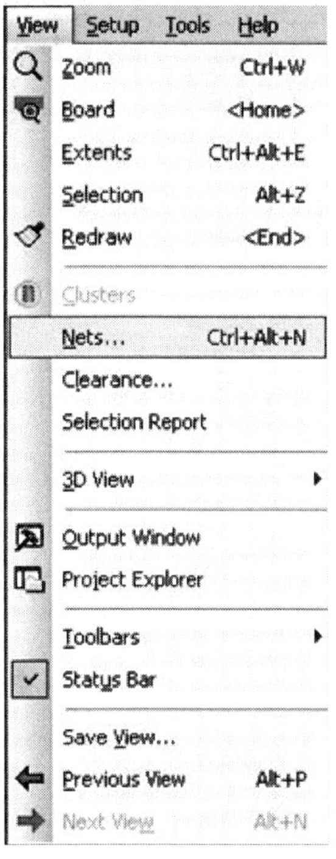

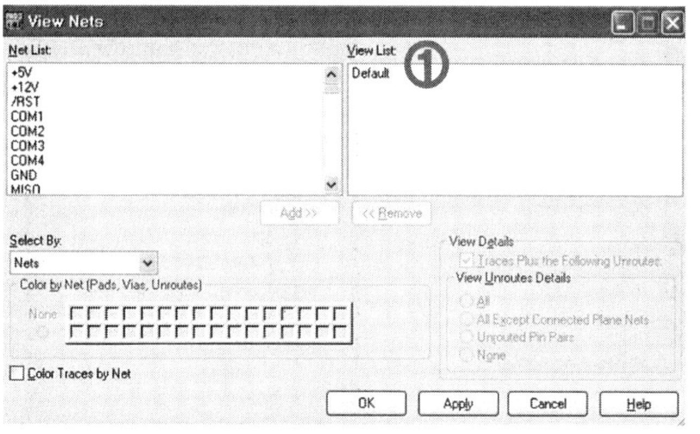

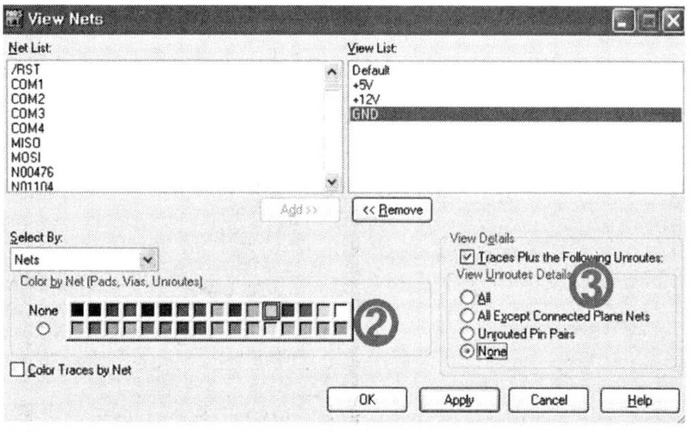

Layout (Nets 활용하기)

가. View – Nets – Net List (특정 Net 선택) – ① View List (Add)

- 특정 Net를 선택하여 Color를 설정할 수 있다.

나. View List (특정 Net 선택) – ② Color by Net 선택 – ③ View Unroutes Details (None 선택)

- 선택된 Net의 Connection을 보이지 않게 할 수 있다. (Ground에 활용)

Layout (Setup 이해)

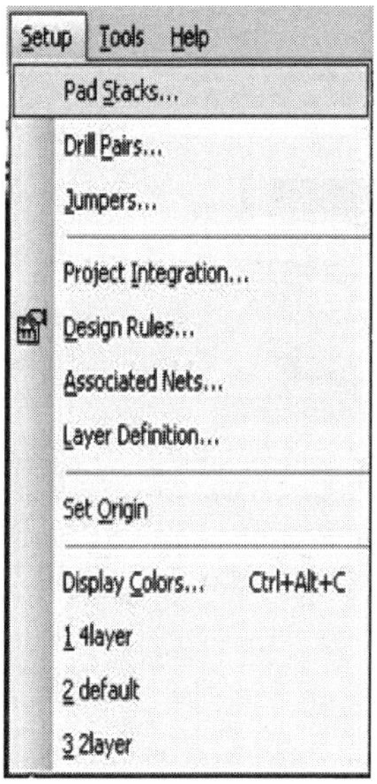

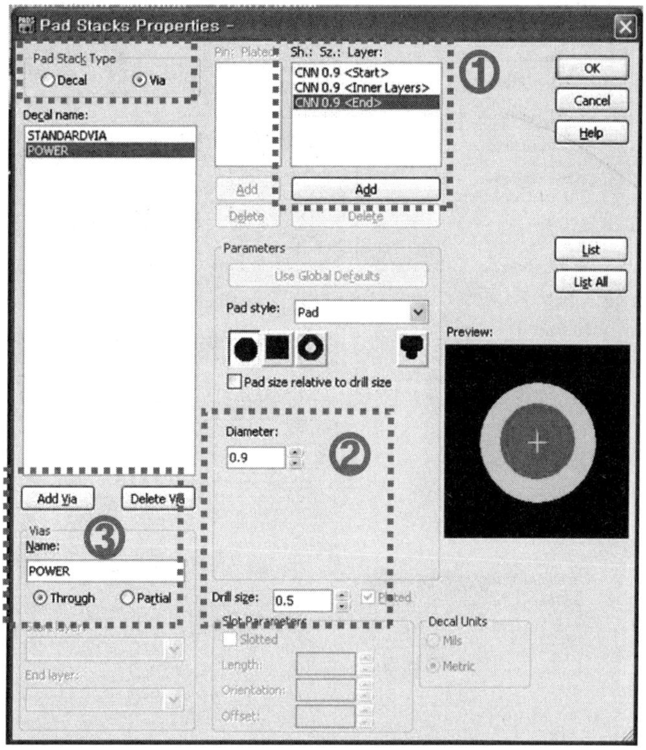

Layout (Via 설정하기)

가. Setup – Pad Stacks – Pad Stack Type (Via 선택)

- Via 속성을 변경하거나 Via를 추가할 수 있다.

나. Decal Name (Standard Via 선택) – ① Sh.: Sz. : Layer: 변경 (Start/Inner/End) – ② Layer별 Diameter (0.6)/Drill size (0.3)

다. ③ Add Via – Name (POWER) – ① Sh.: Sz. : Layer: 변경 (Start/Inner/End) ② Layer별 Diameter (0.9)/Drill size (0.5) – OK

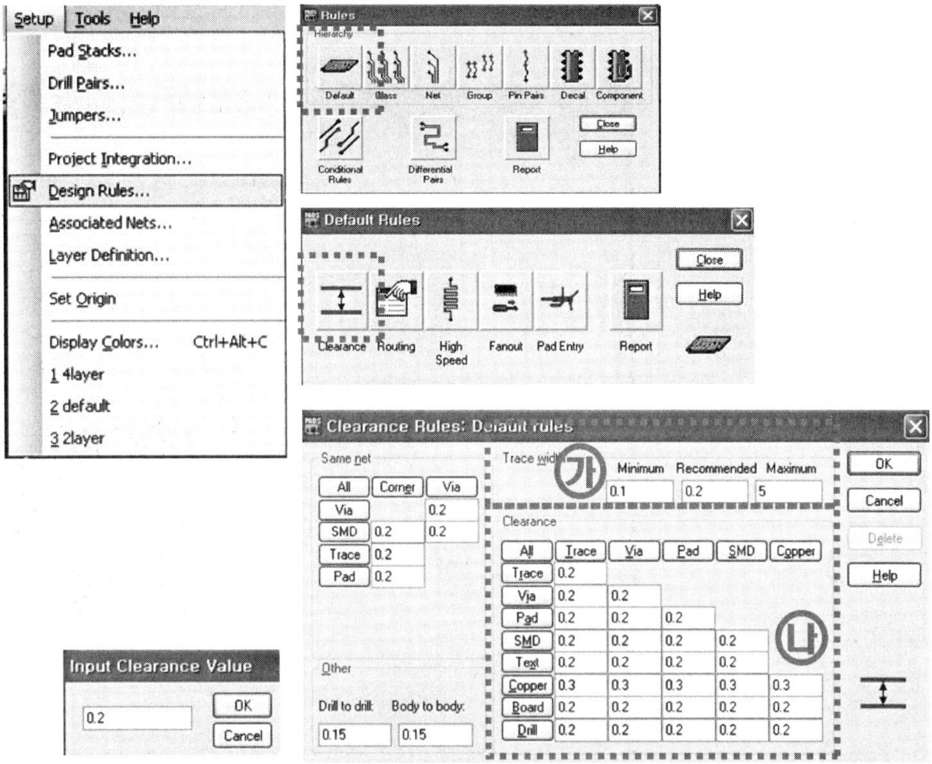

Layout (Design Rules 설정하기)

가. Setup – Design Rules – Rules (Default 선택) – Default Rules (Clearance 선택) – Clearance Rules – Trace width (Min0.1/Rec0.2(배선두께)/Max5)

- 기본 배선 두께를 설정한다. (Recommended=기본 배선 두께)

나. Setup – Design Rules – Rules (Default 선택) – Default Rules (Clearance 선택) – Clearance Rules – Clearance (ALL0.2/Copper0.3) – OK – OK

- 모든 Item간의 이격 거리를 설정할 수 있다.

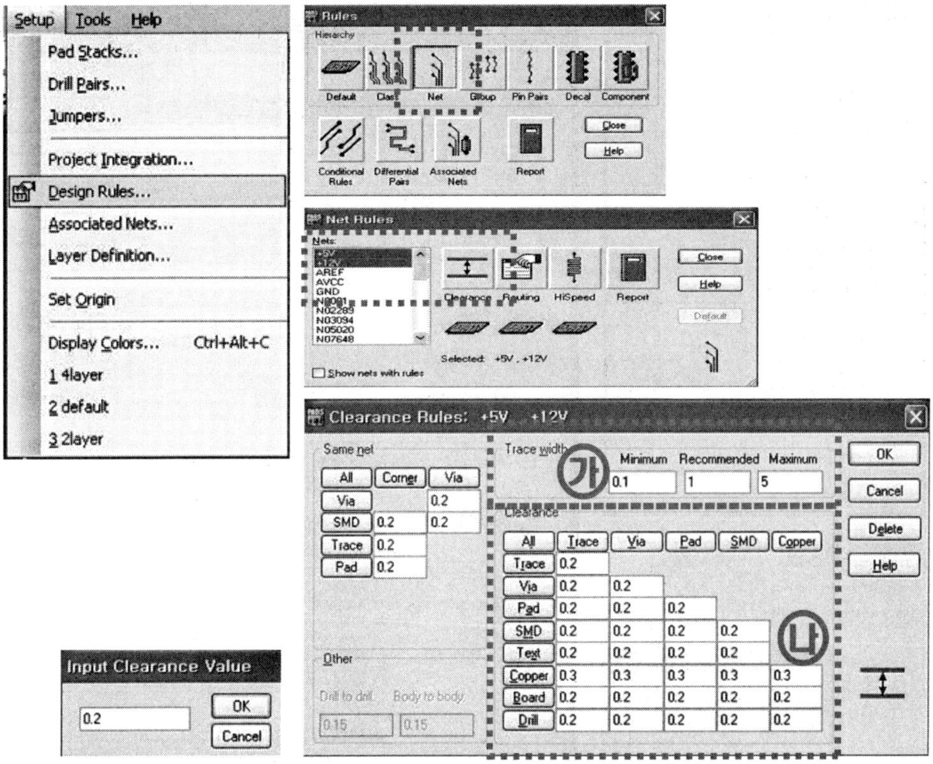

Layout (Design Rules 설정하기)

가. Setup – Design Rules – Rules (Net 선택) – Net Rules (특정 Net 선택) – Clearance 선택 – Clearance Rules – Trace width (0.1/1(배선두께)/5)

- 특정 Net 배선 두께를 설정한다.(Recommended= 특정 Net 배선 두께)

나. Setup – Design Rules – Rules (Net 선택) – Net Rules (특정 Net 선택) – Clearance 선택 – Clearance Rules – Clearance (ALL0.2/Copper0.3) – OK – OK

- 특정 Net의 모든 Item간의 이격 거리를 설정할 수 있다.

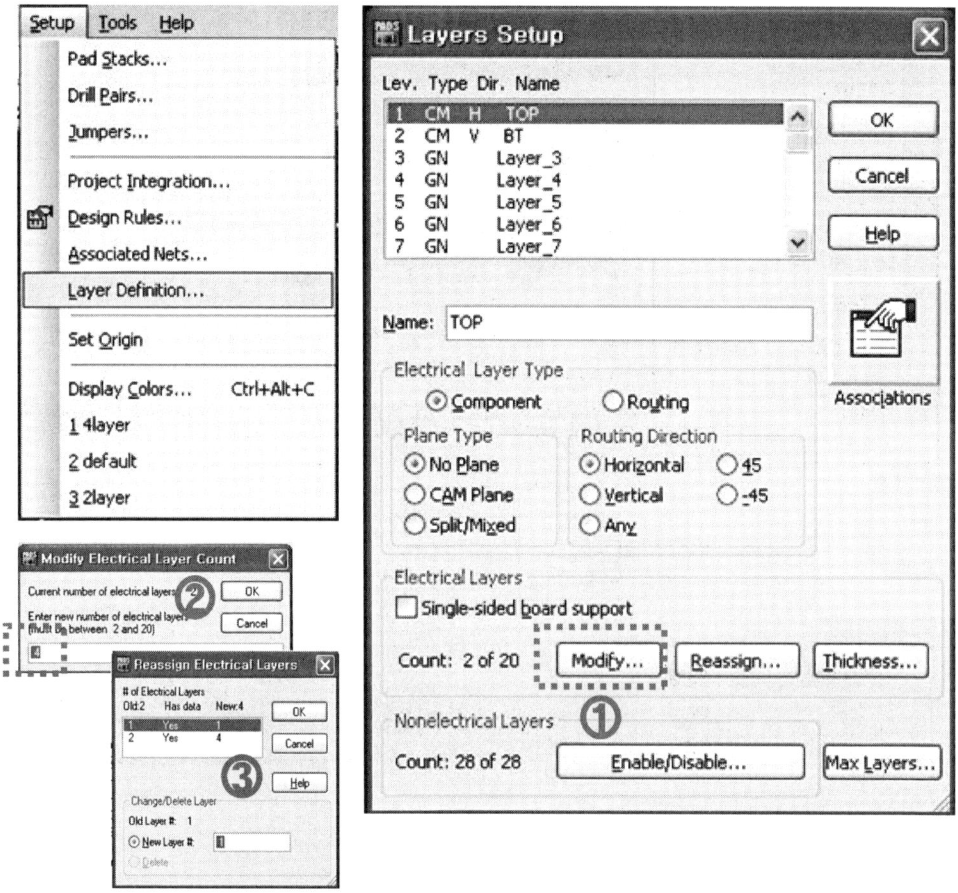

Layout (Layer Definition 설정하기)

가. Setup – Layer Definition – Layer Setup – ① Electrical Layer (Modify 선택) – ② Modify Electrical Layer Count (4) – OK – ③ Reassign Electrical Layers – OK

- Layer를 추가 및 제거할 수 있다.

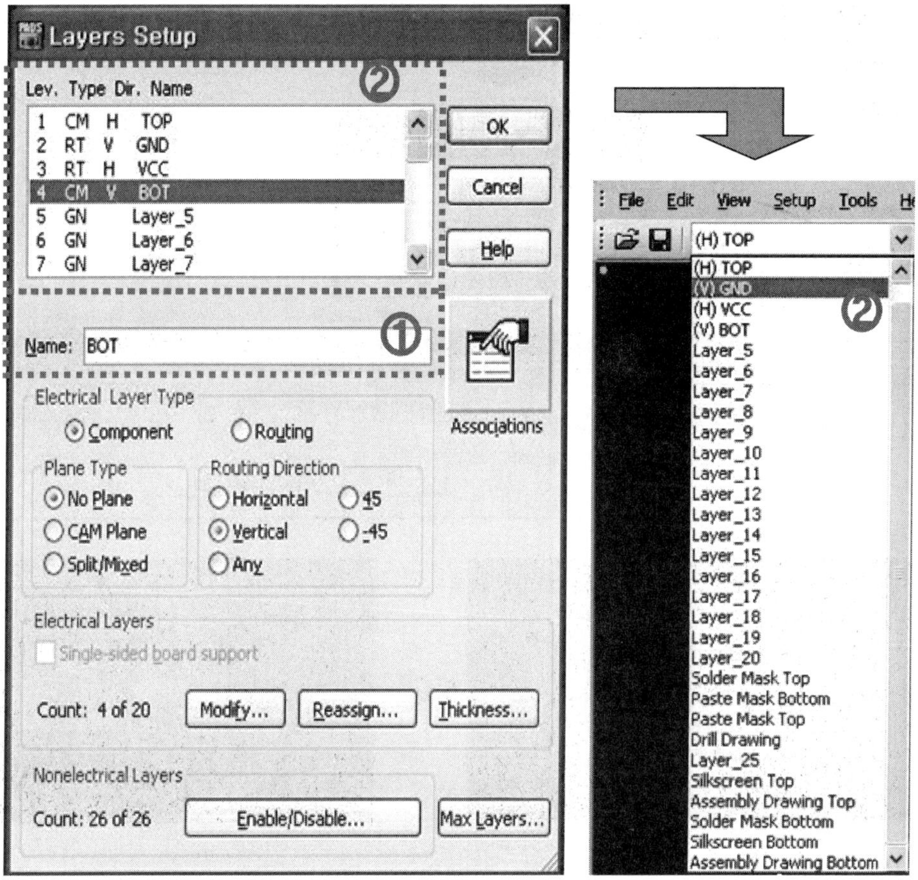

Layout (4Layer Definition 설정하기)

가. Layer Setup – 4 Layer – ② Lev. Type Dir. Name (Layer 선택) – ① Name (BOT)

- 각 Layer의 Name을 변경할 수 있다.
- 변경된 Layer는 Main화면에서 확인할 수 있다.

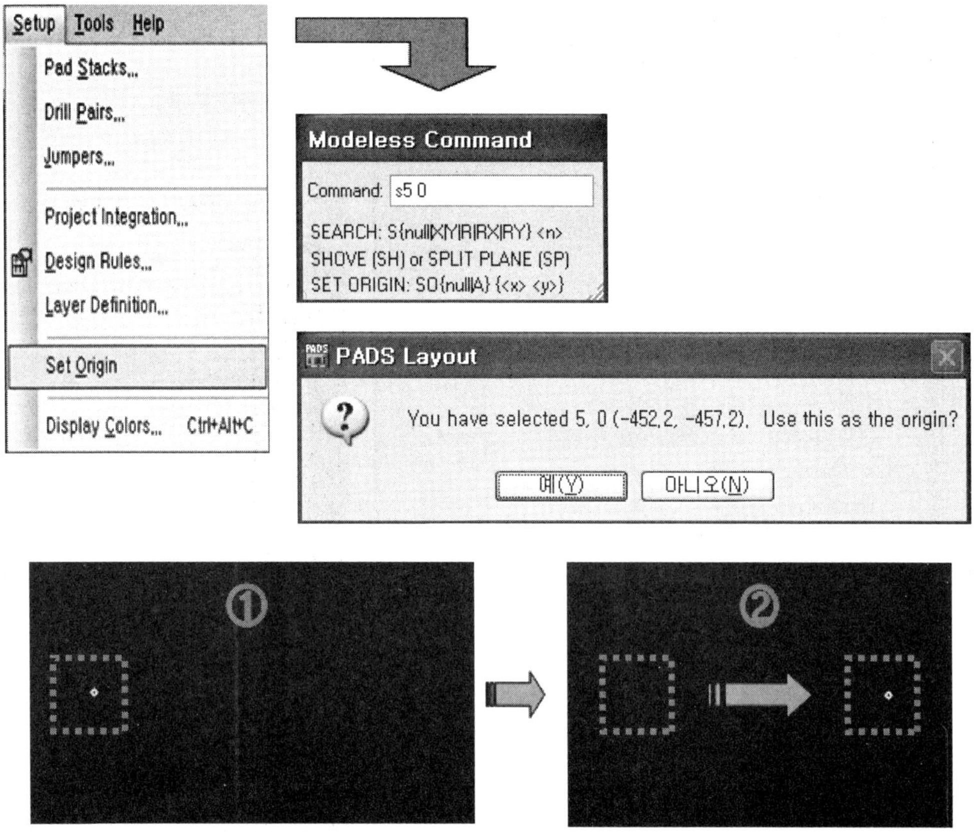

Layout (Set Origin 설정하기)

가. Setup -Set Origin -S(X좌표 Y좌표)-Enter -Space bar -Pads Layout(예) -Yes

- 그림과 같이 ①번-〉 ②번으로 이동하여 그림과 같이 원점을 변경한다.

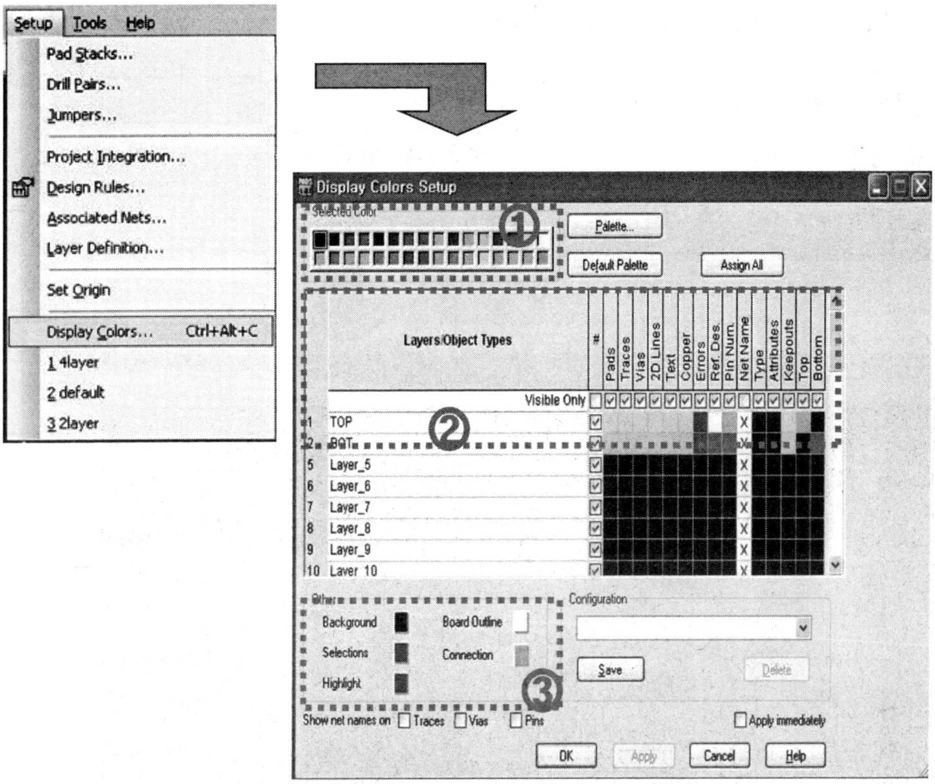

Layout (Display Color 설정하기)

가. Setup – Display Colors – ① Select Color 선택 – ② Layer 별 특정 Item 선택(Color 변경) – ③ Other 확인 – OK

- Layout상에 Layer를 Color로 구분한다.

Layout (PCB Decal Editor)

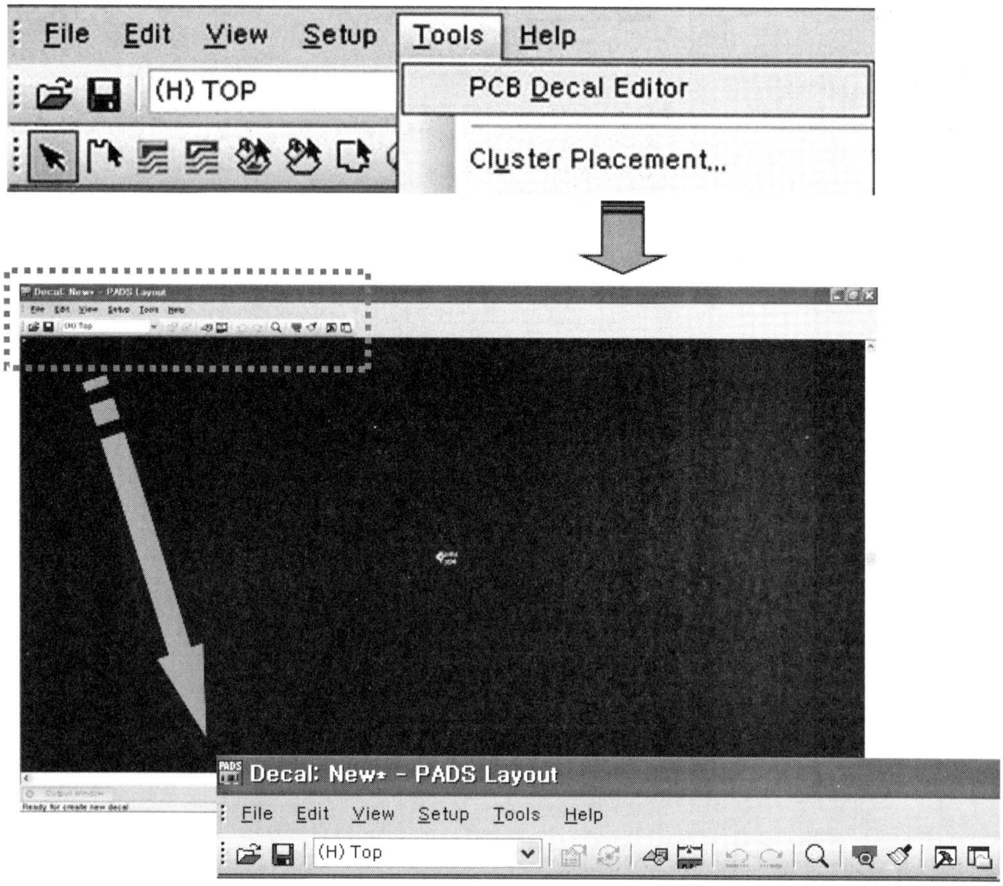

Layout (PCB Decal Edotor)

가. Tools -PCB Decal Editor

- 기본 창에서 Library로 들어가 Decal작성을 한다.
- Library화면에서는 부품의 Name과 Type이 원점에 표시된다.

Layout (Disperse Components)

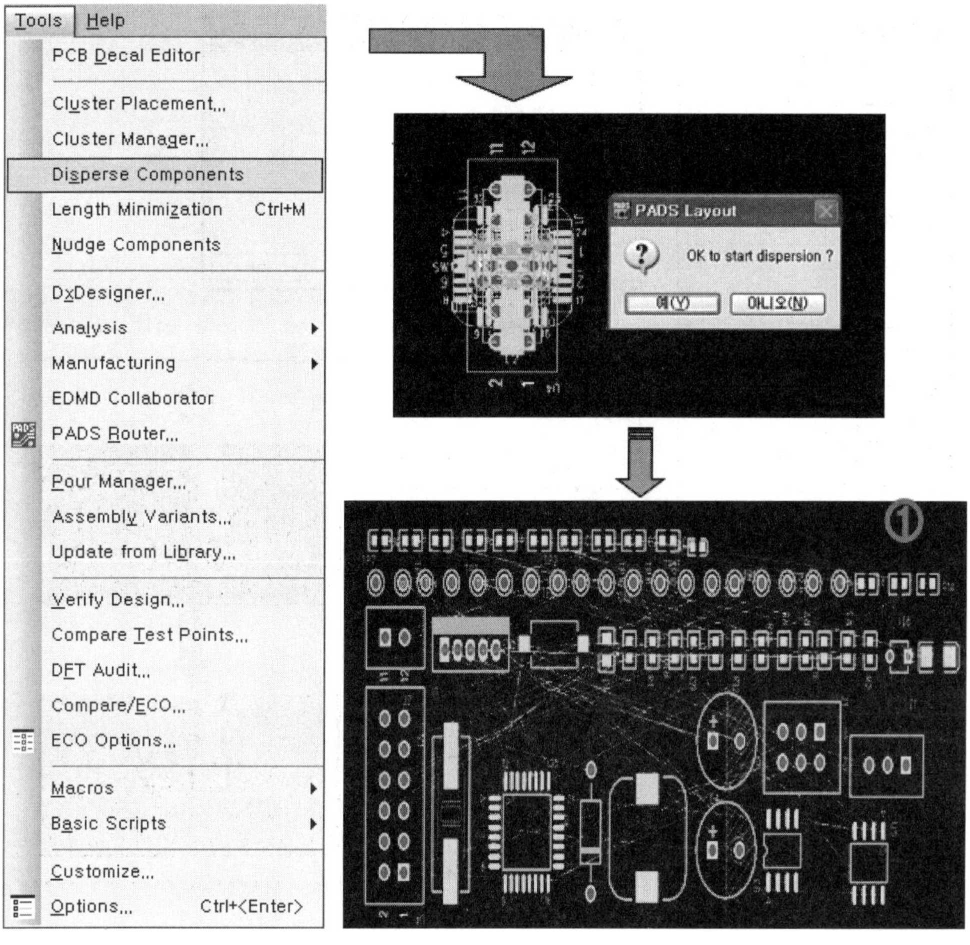

Layout (Disperse Components)

가. Tools -Disperse Components -PADS Layout -예

- ①의 그림과 같이 원점에 겹쳐있던 부품을 펼쳐 보일 수 있다.

Layout (Pour Manager)

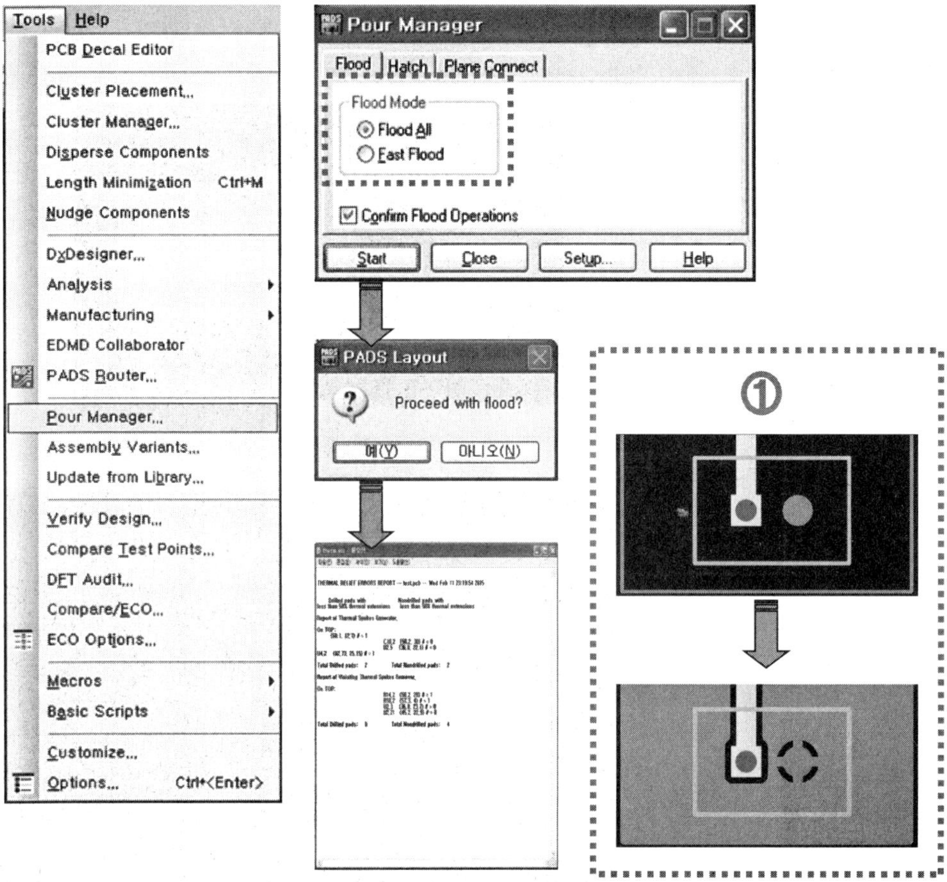

Layout (Pour Manager)

가. Tools – Pour Manager – Flood All 선택 – Start – PADS Layout (예) – 창 끄기

- ①의 그림과 같이 Copper Pour Outline을 기준으로 채워진다.

Layout (Error 이해)

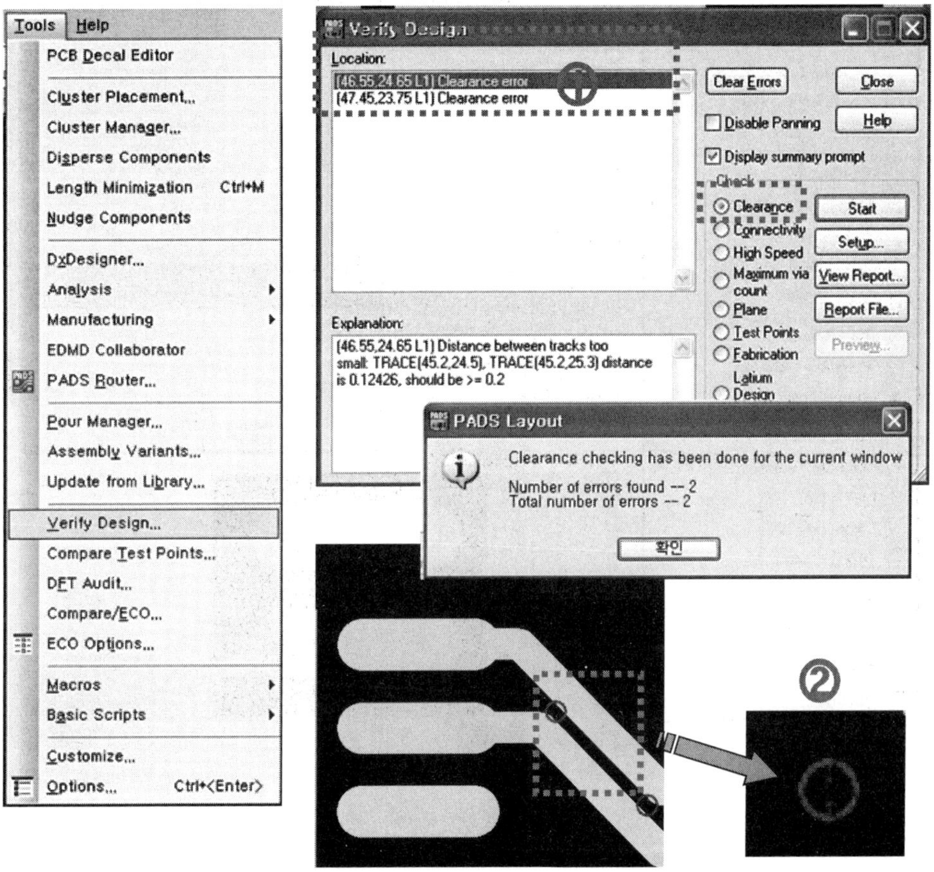

Layout (Verify Design)

가. Tools – Display Colors – Verify Design – Clearance 선택 – Start – PADS Layout (확인) – ① Location – ② Clearance Error 확인

- 이격 거리가 설정보다 가까울 경우 Error를 찾아낸다.

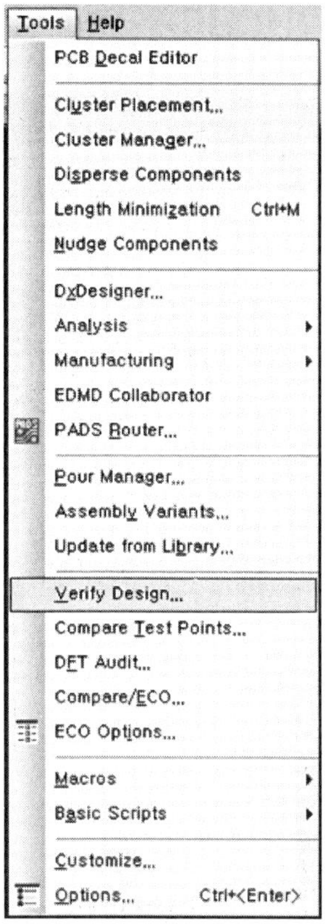

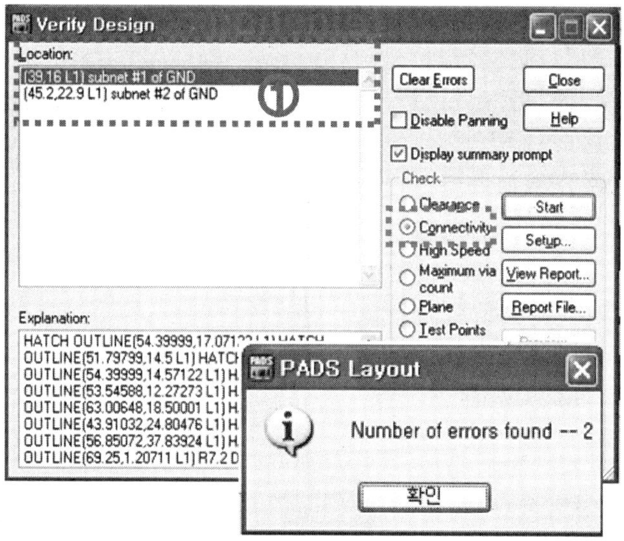

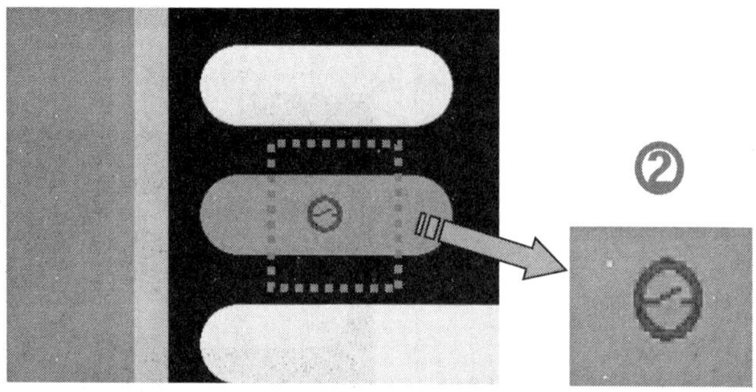

Layout (Verify Design)

가. Tools – Display Colors – Verify Design – Connectivity 선택 – Start – PADS Layout (확인) – ① Location – ② Connectivity Error 확인

- Connection이 연결되어 있지만 배선이 되지 않을 경우 Error를 찾아낸다.

Layout (Basic Scripts)

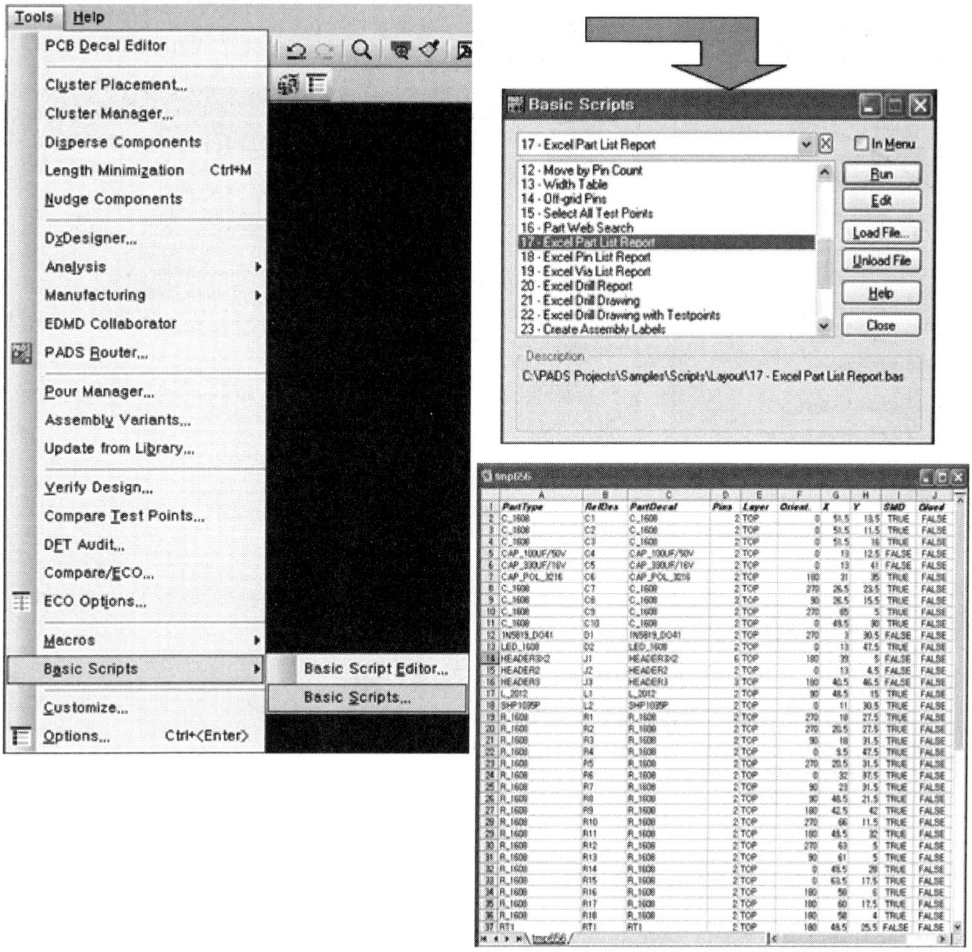

Layout (Basic Scripts)

가. Tool -Basic Scripts -Basic Scripts -17번 선택 -Run -Excel(좌표 데이터)

- 부품의 좌표데이터를 엑셀 파일로 생성할 수 있다.

Layout (Option 이해)

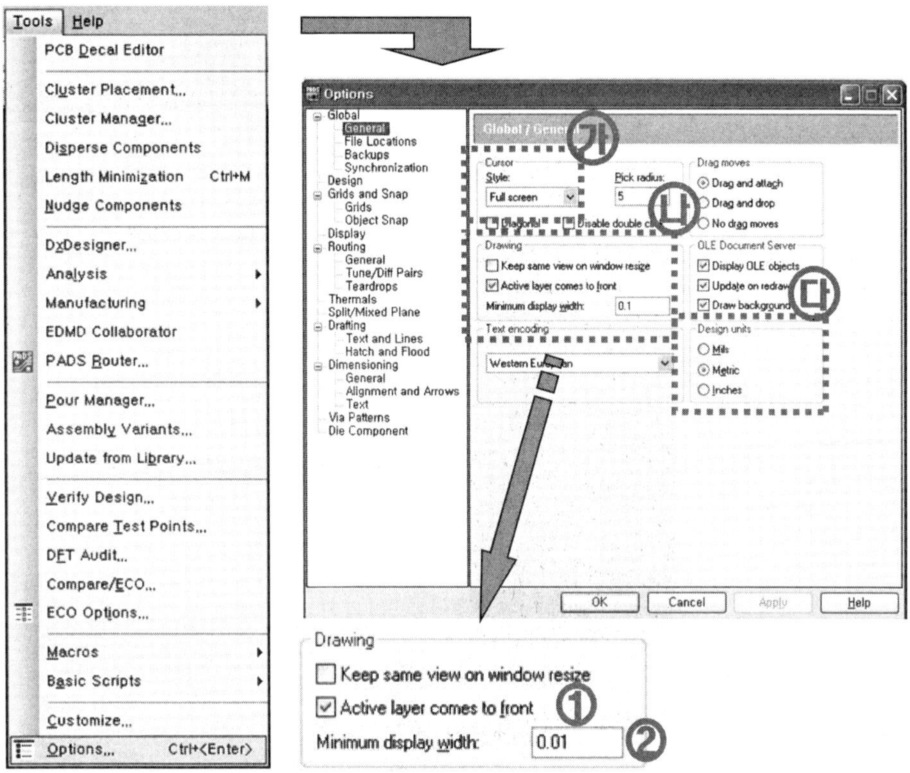

Layout (Option - Global)

가. Tools – Options – Global – General – Cursor – Style (4가지 중 선택)
– Cursor의 모양을 정할 수 있다. (Normal, Small, Large, Full screen)

나. Tools – Options – Global – General – Drawing – Active Layer/Min Display (0.01)

- ① Layer 위치 변경 시 Display 된다. / ② 디스플레이 되는 두께의 최소 치수이다.

다. Tools – Options – Global – General – Design units 선택

- Decal 작성시 단위 설정을 한다. (Mils, Metric (mm), Inches)

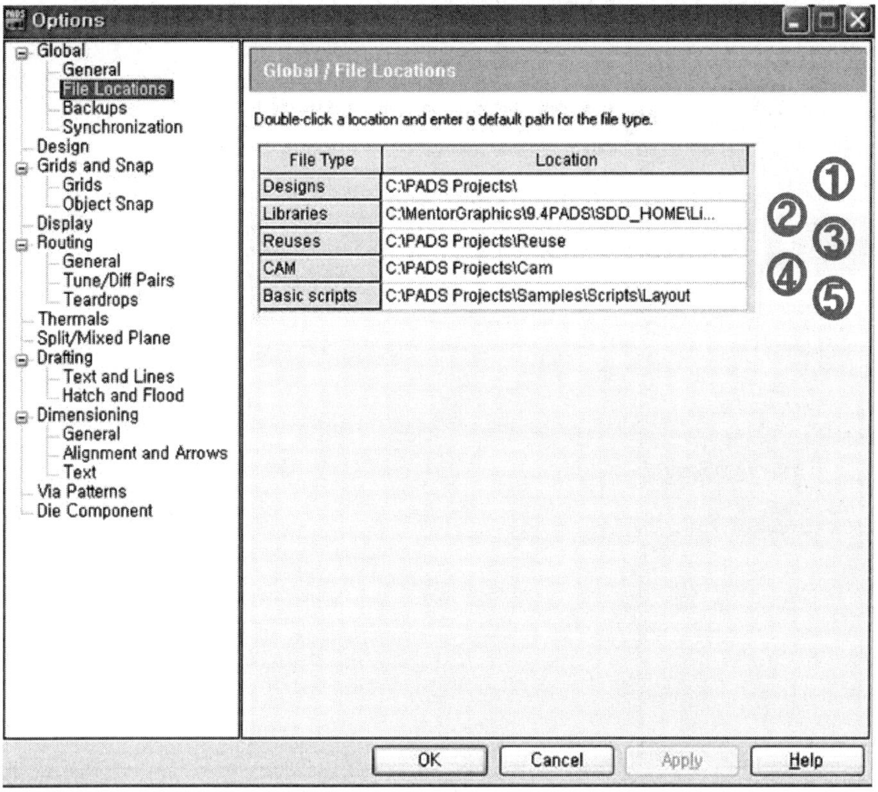

Layout (Option - File Location)

가. File 저장경로

① Tools - Options - Global - File Location - Designs 저장경로 (PADS File)

② Tools - Options - Global - File Location - Libraries 저장경로 (Library)

③ Tools - Options - Global - File Location - Reuses 저장경로 (Reuse)

④ Tools - Options - Global - File Location - CAM 저장경로 (Gerber)

⑤ Tools - Options - Global - File Location - Basic Scripts 저장경로 (좌표data)

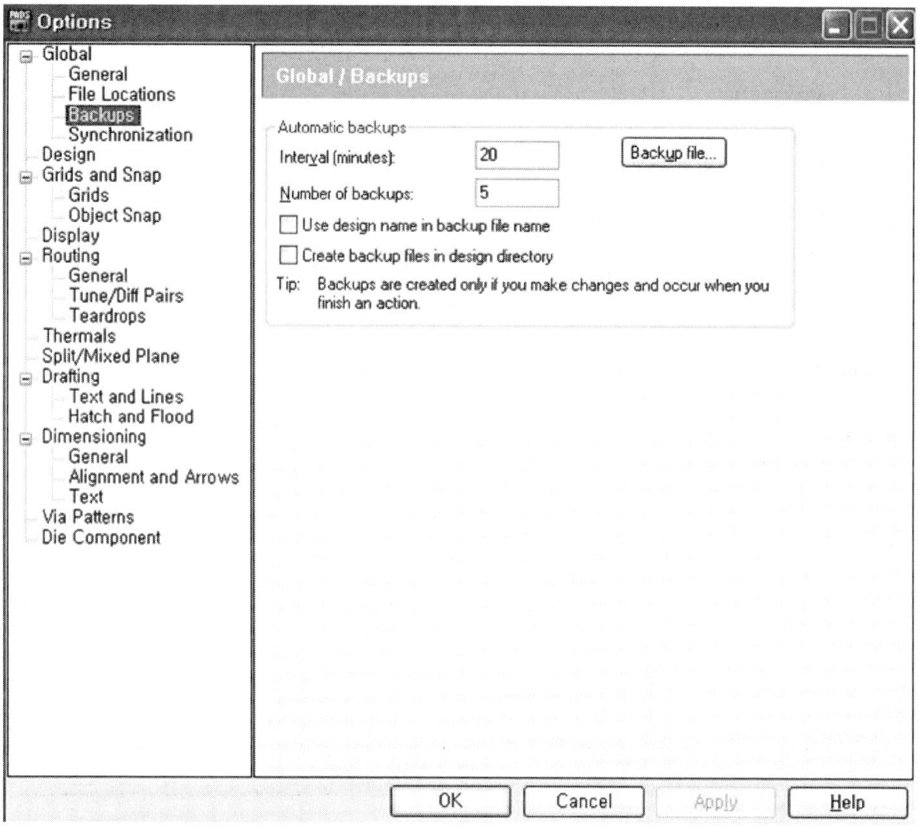

Layout (Option - Backups)

가. Tools – Options – Global – Backups – Automatic backups – Interval – 20

- 20분 간격으로 Backup File이 생성된다.

나. Tools – Options – Global – Backups – Automatic backups – Number of – 5

- 5개의 Backup File이 자동 저장된다.

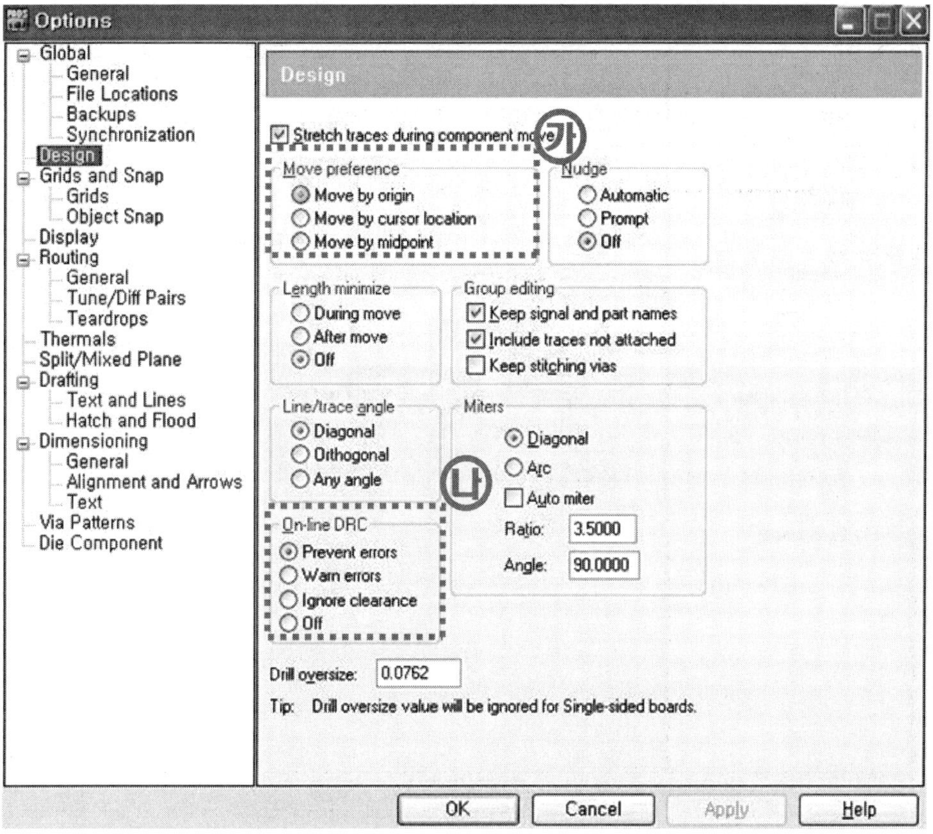

Layout (Option - Design)

가. Tools – Options – Design – Move preference – 3개 중 선택
- 객체를 선택할 경우 Cursor의 위치를 선택할 수 있다.

나. Tools – Options – Design – On-line DRC – Prevent errors or Off
- 배선 시 이격 거리를 유지하며 Routing 할 수 있다.

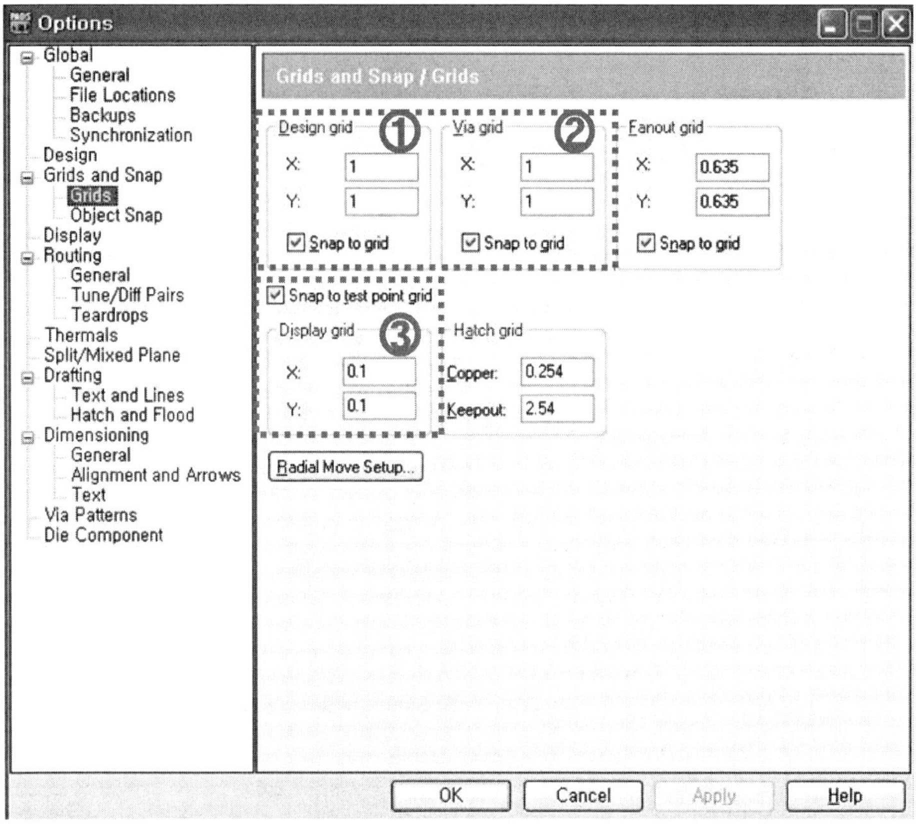

Layout (Option - Grid and Snap)

가. Tools – Options – Grid and Snap – Grids – ① Design / ② Via Grid (1,1/ 1,1)

- ① Cursor의 움직임의 최소단위 / ② Via 생성 시 최소단위

나. Tools – Options – Grid and Snap – Grids – ③ Display Grid(0.1, 0.1)

- Background에 Display 되는 Point의 간격

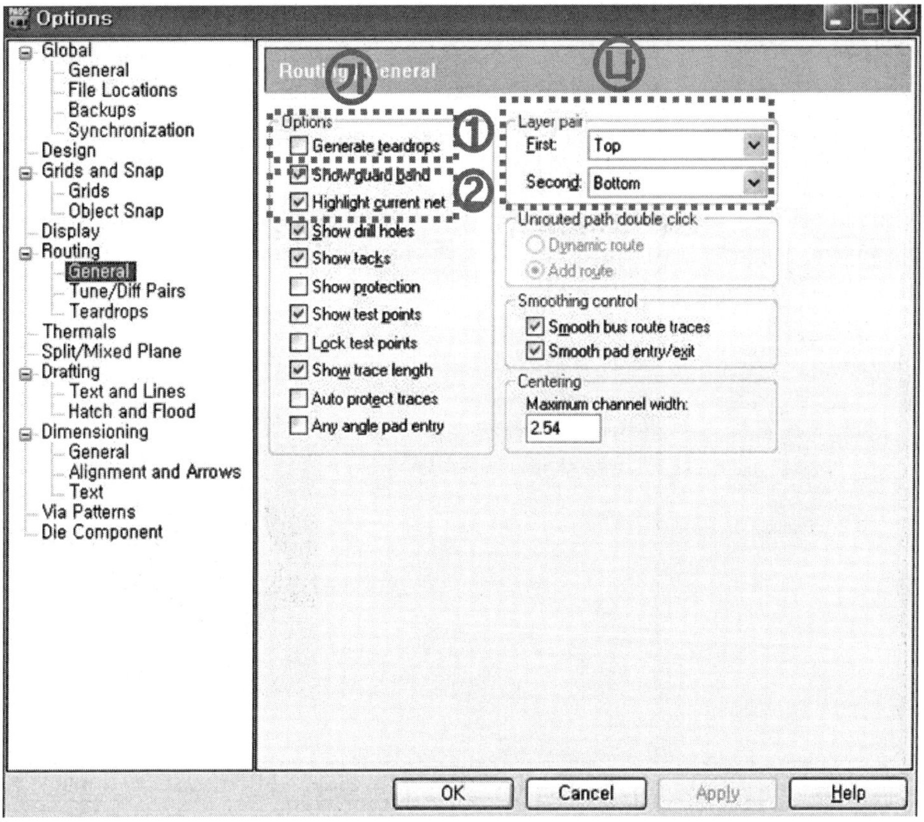

Layout (Option - Routing)

가. Tools – Options – Routing – General – ① Teardrops / ② Highlight current net

- ① Pin에서 배선 시 Teardrop 생성 / ② 배선 시 net를 활성화시킨다.

나. Tools – Options – Routing – General – Layer pair (First/Second)

- 자주 사용하는 Layer별로 설정 시 쉽게 이동이 가능하다.

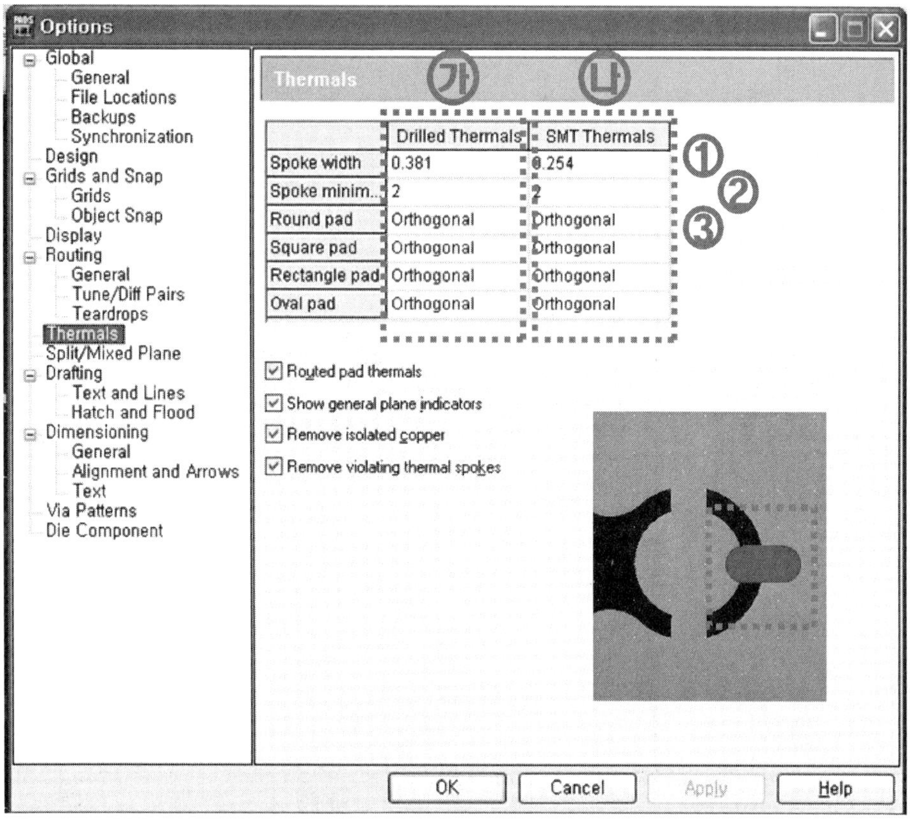

Layout (Option - Thermals)

가. Tools – Options – Thermals – Drill – ① Spoke Width (두께) ② Spoke Min. (개수) ③ Round/Square/Rectangle/Oval Pad (Pad 모양에 따른 Spoke 방향설정)

- Copper Pour 작업 시 생성되는 Spoke의 설정을 변경할 수 있다.

나. Tools – Options – Thermals – SMT – ①②③ 설정 (SMD PAD에 적용)

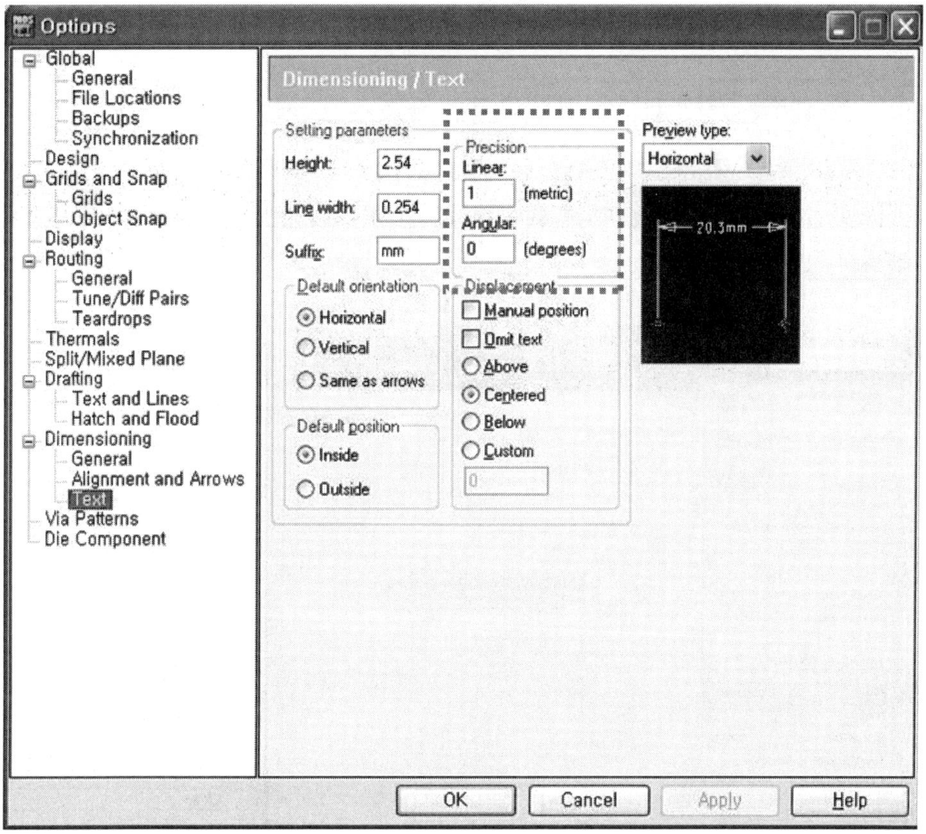

Layout (Option - Text)

가. Tools – Options – Dimensioning – Text – Precision – Linear (2)

- 치수를 표시하는 Dimensioning의 소수점 자릿수이다.

Layout 이해(Drafting Toolbar)

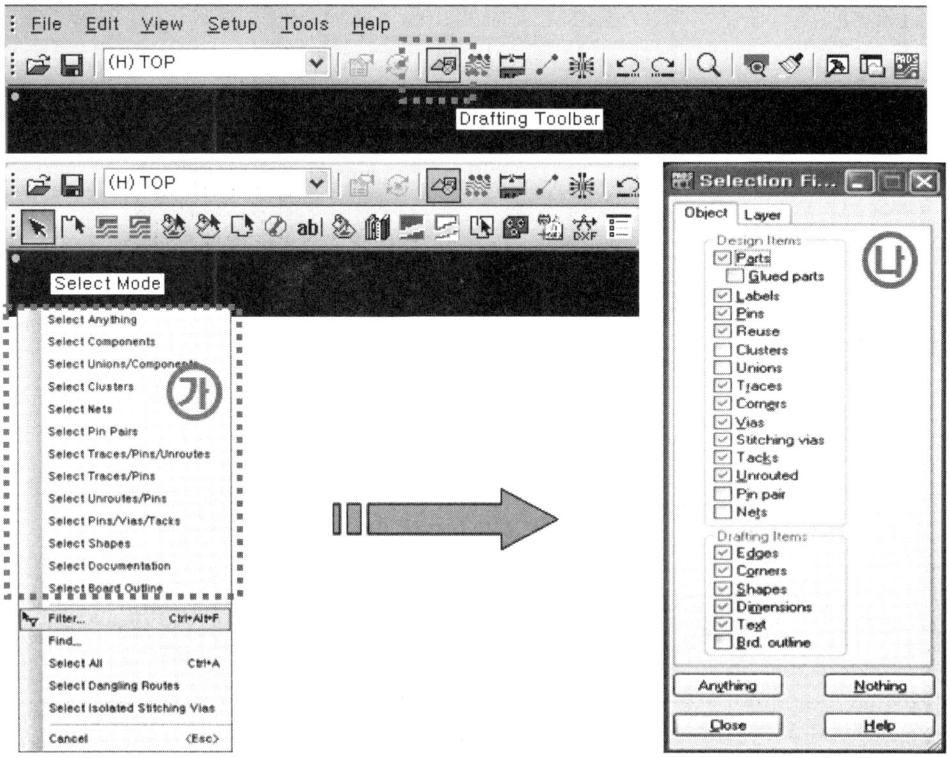

Select Mode

가. Drafting Toolbar – Select -마우스(우측) *단.선택된 Item 없을 경우

- 자주 쓰이는 Items을 선택할 수 있다.

나. Drafting Toolbar – Select -마우스(우측) – Filter - Anything or Nothing

- Design Items (Part=부품, Glued parts=고정부품, Labels=부품번호, Traces=패턴, Vias=층간 홀, Stitching vias=층간 고정 홀, Pin pair=네트의 일부, Nets=네트의 전체)
- Drafting Items (Library Filter와 동일, Brd. Outline=PCB 외각)

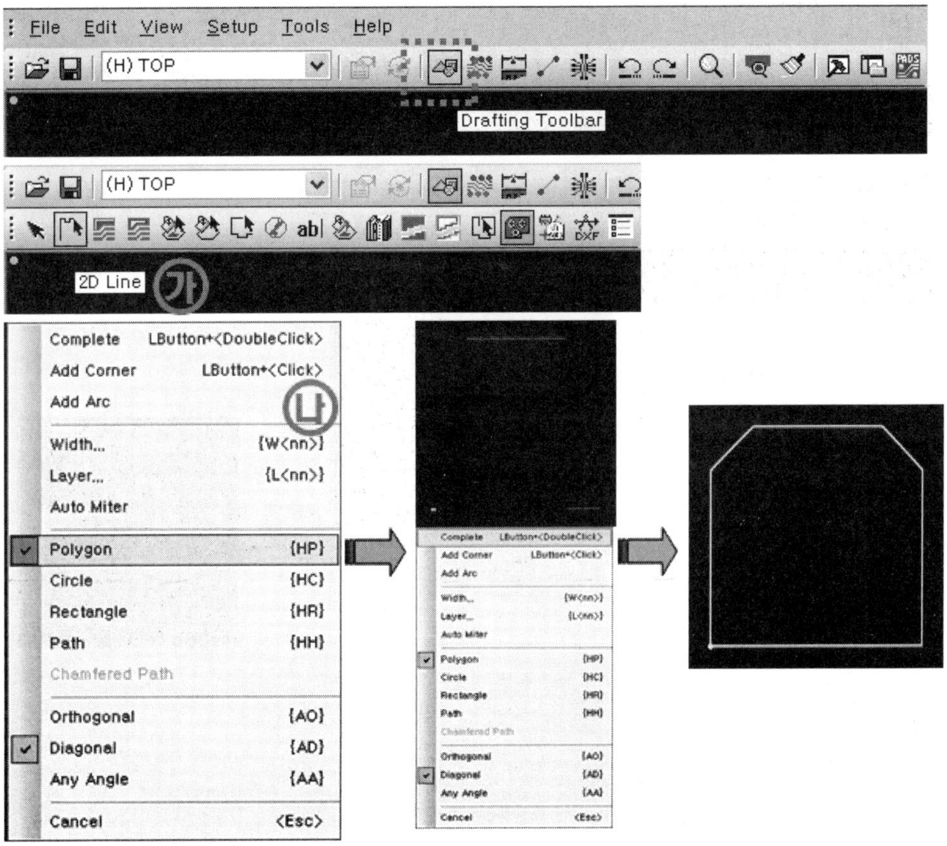

2D Line Mode (Polygon)

가. Drafting Toolbar -2D Line

- PCB의 외형이나 도형을 생성할 수 있다.

나. Drafting Toolbar – 2D Line -마우스(우측) -Polygon -도형생성 -마우스(우측) – Complete

- Polygon : 선으로 그린 도형을 box로 완성할 수 있다.

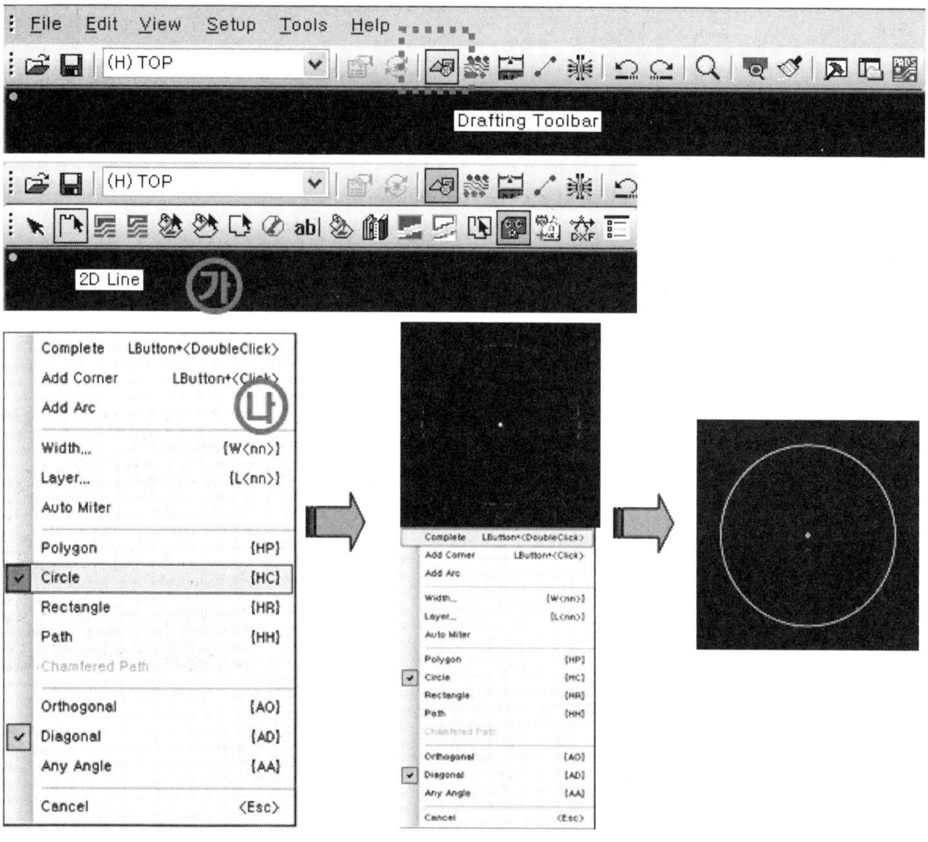

2D Line Mode (Circle)

가. Drafting Toolbar -2D Line

- PCB의 외형이나 도형을 생성할 수 있다.

나. Drafting Toolbar – 2D Line -마우스(우측) -Circle -도형생성 – 마우스(우측) – Complete

- Circle : 원형의 도형을 완성할 수 있다.

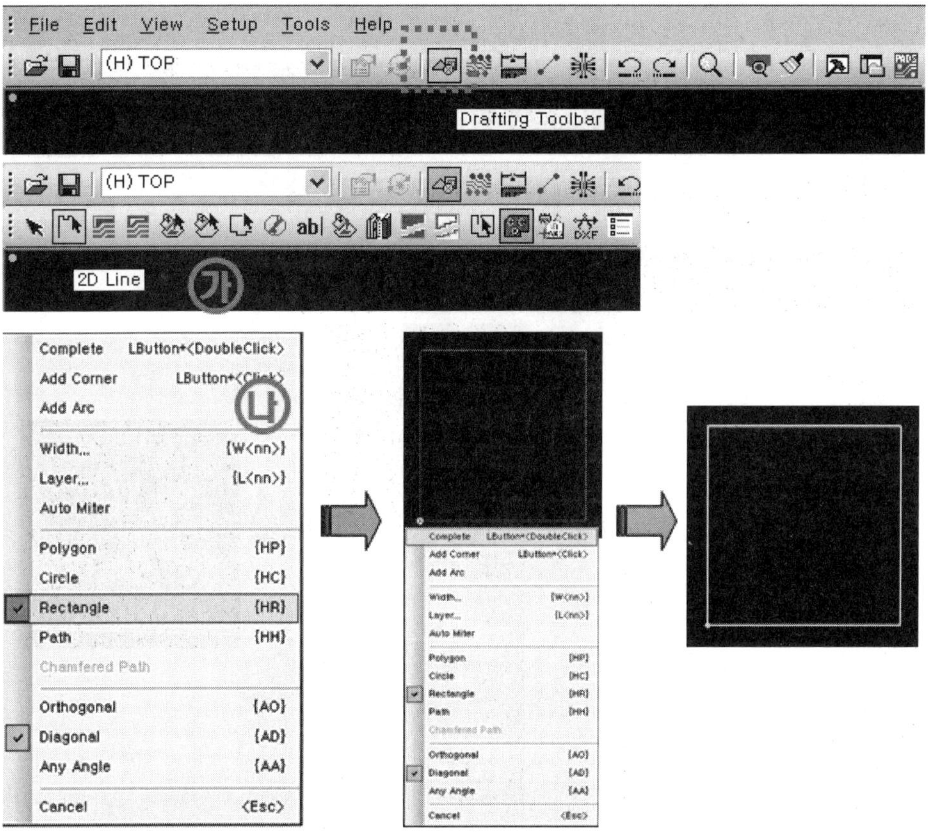

2D Line Mode (Rectangle)

가. Drafting Toolbar -2D Line

- PCB의 외형이나 도형을 생성할 수 있다.

나. Drafting Toolbar – 2D Line -마우스(우측) -Rectangle -도형생성 -마우스(우측) – Complete

- Rectangle : Box의 도형을 완성할 수 있다.

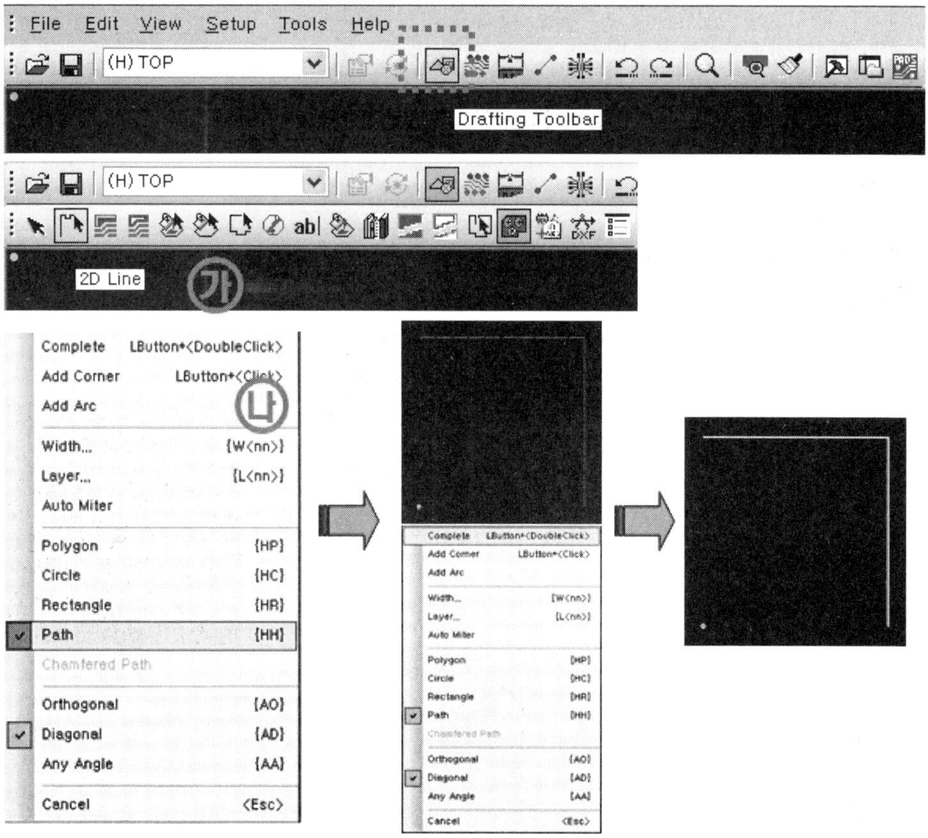

2D Line Mode (Path)

가. Drafting Toolbar -2D Line

- PCB의 외형이나 도형을 생성할 수 있다.

나. Drafting Toolbar – 2D Line -마우스(우측) -Path -도형생성 -마우스(우측) -Complete

- Path : 선의 도형을 완성할 수 있다.

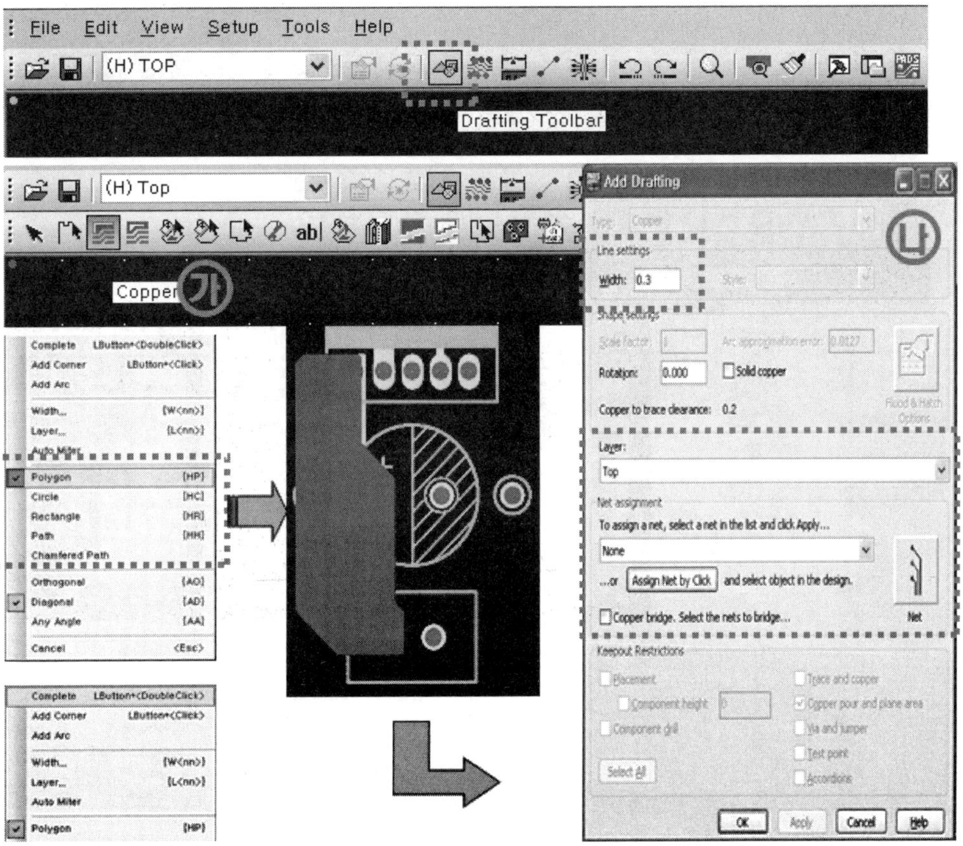

Copper Mode

가. Drafting Toolbar -Copper

- 전원선과 같이 보강된 패턴을 그릴 수 있다. (5가지 도형생성 중 선택)

나. Drafting Toolbar – Copper -마우스(우측) -Polygon -도형생성 -마우스(우측) -Complete – Add Drafting (Width=도형의 선 두께, Layer=생성 층, Net assignment=도형의 net속성 변경) -OK

- PCB의 보강패턴 생성이나 실크추가에 활용할 수 있다.

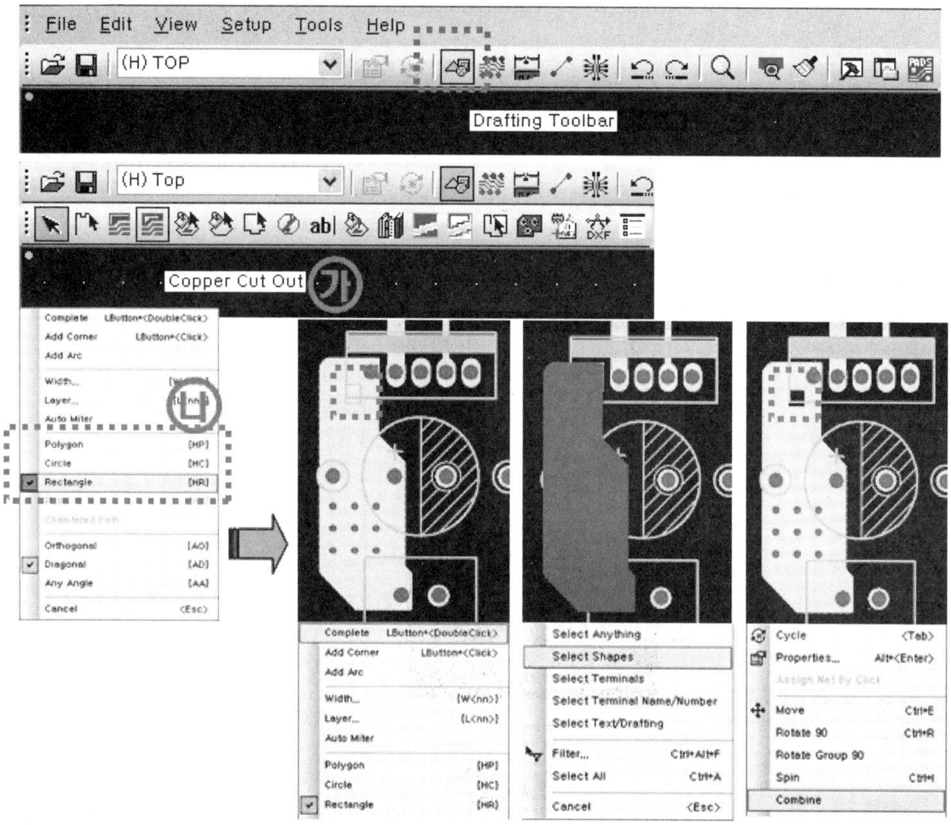

Copper Cut Out Mode

가. Drafting Toolbar -Copper Cut Out

- Filled된 도형을 잘라 낼 수 있다. (Polygon, Circle, Rectangle)

나. Drafting Toolbar – Copper Cut Out-마우스(우측) -Rectangle -도형생성 -Select Mode -마우스(우측) -Select Shapes -전체도형 선택 – 마우스(우측) -Combine

- PCB의 보강패턴이나 실크의 변경에 활용할 수 있다.

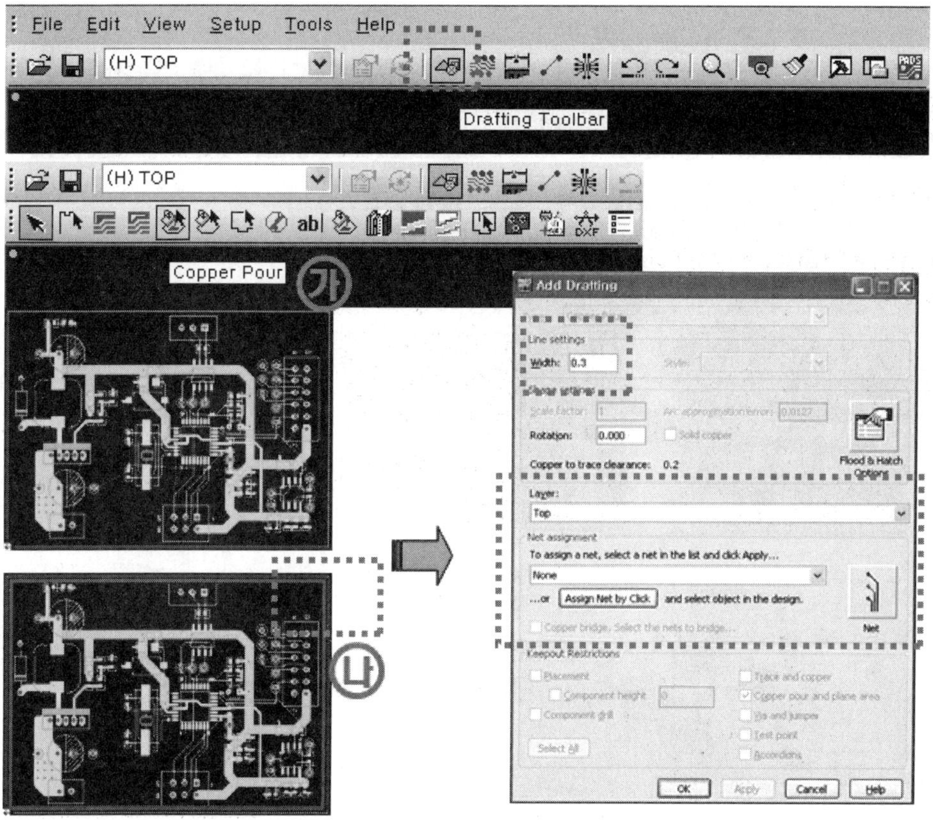

Copper Pour Mode

가. Drafting Toolbar -Copper Pour

- GND 패턴을 연결하기 위해 Copper pour의 외각라인을 그린다. (3가지 유형)

나. Drafting Toolbar – Copper Pour-마우스(우측) -Polygon -도형생성 -마우스(우측) -Complete – Add Drafting (Width=도형의 선 두께, Layer=생성 층, Net assignment=도형의 net속성 변경) -OK

- Copper Pour를 이용하여 빈 공간에 Ground의 배선영역을 생성할 수 있다.

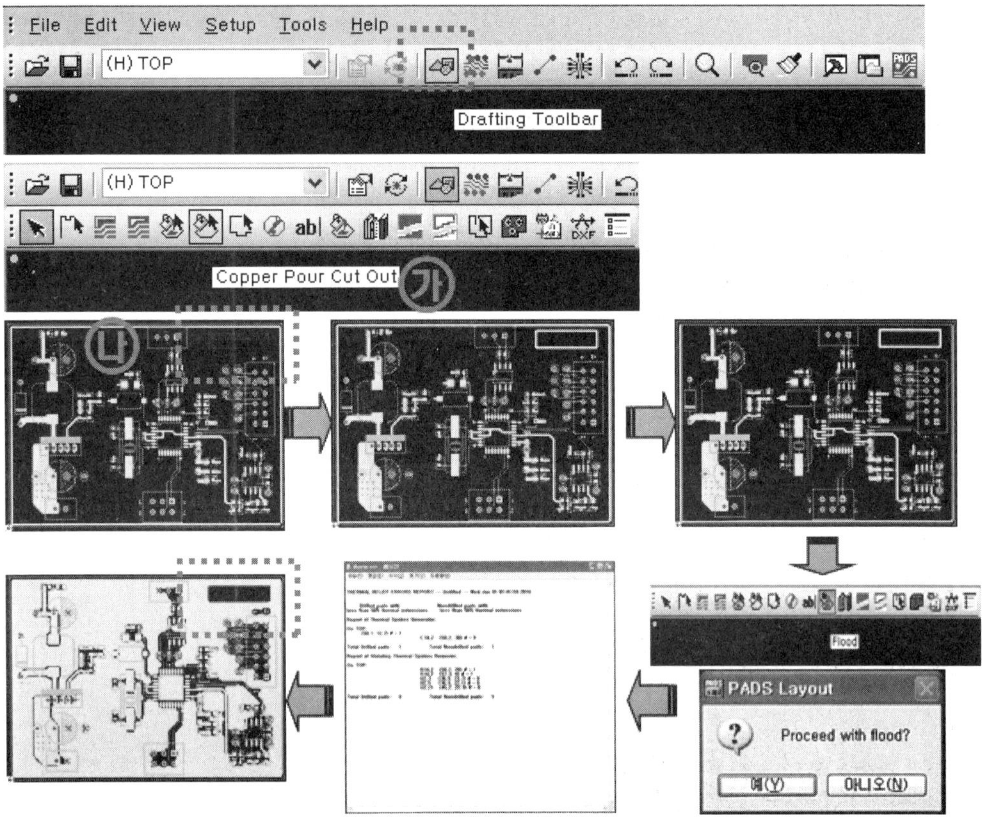

Copper Pour Cut out Mode

가. Drafting Toolbar -Copper Pour Cut Out

- Copper pour의 일부분을 오려낼 수 있다. (Polygon, Circle, Rectangle)

나. Drafting Toolbar – Copper Pour Cut Out-마우스(우측) -Rectangle -도형생성 -Select Mode -Copper Pour 선택 – Flood -PADS Layout -예 -Therm. Err 닫기

- Ground의 배선영역에서 Copper Pour를 삭제 또는 변경할 수 있다.

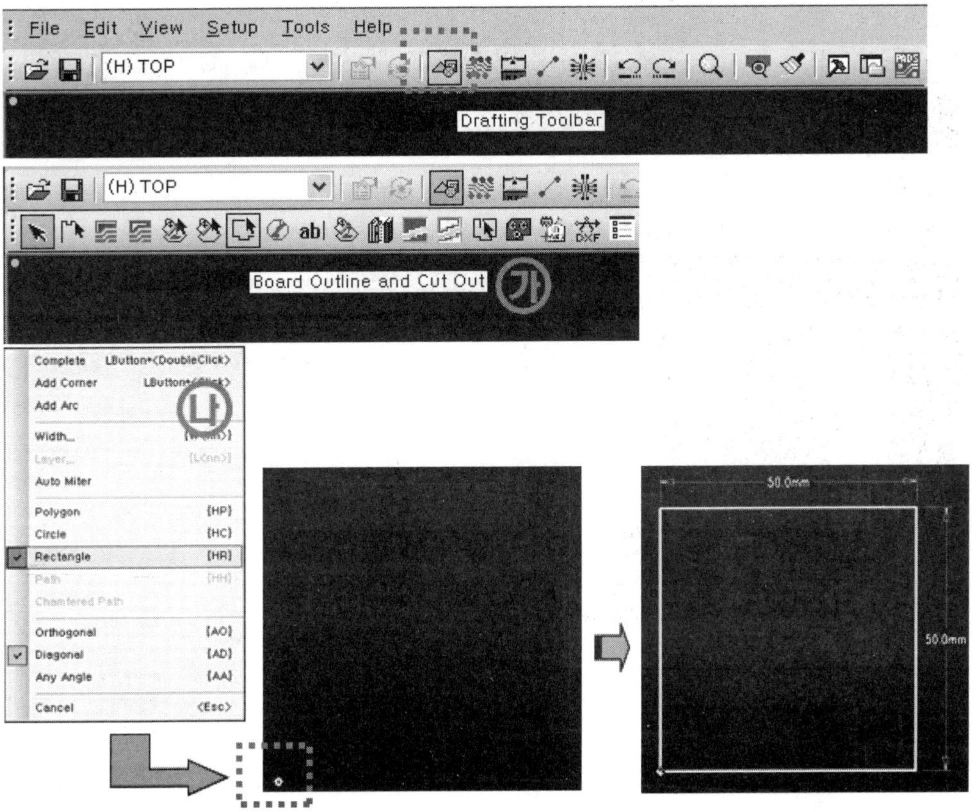

Board Outline and Cut out Mode

가. Drafting Toolbar -Board Outline and Cut Out

- PCB의 외형을 생성할 수 있다. (Polygon, Circle, Rectangle)

나. Drafting Toolbar -Board Outline and Cut Out 선택 -마우스(우측) -Rectangle – 원점 이동 S⟨x, y⟩ -Enter -Space –외형 size 이동 -Enter – Space

- 기준점을 좌측 하단으로 정해 PCB 외형을 그린다.

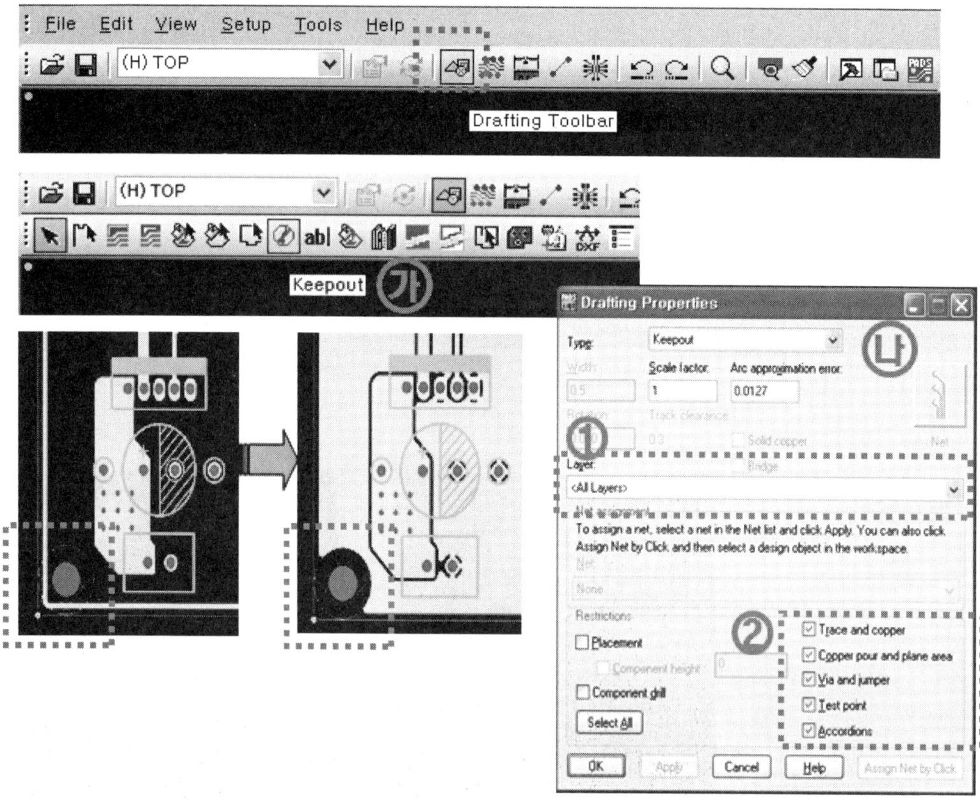

Keep Out Mode

가. Drafting Toolbar -Keep Out

- 금지구역을 설정 할 수 있다. (Polygon, Circle, Rectangle)

나. Drafting Toolbar -Keep Out 선택 -마우스(우측) -Circle – 위치이동 S〈x, y〉 -도형생성 -Drafting Properties -① Layer(생성 층) -② restrictions(금지항목) 선택 – OK

- 레이어 별로 금지항목을 선택하여 금지구역을 설정해 놓을 수 있다.

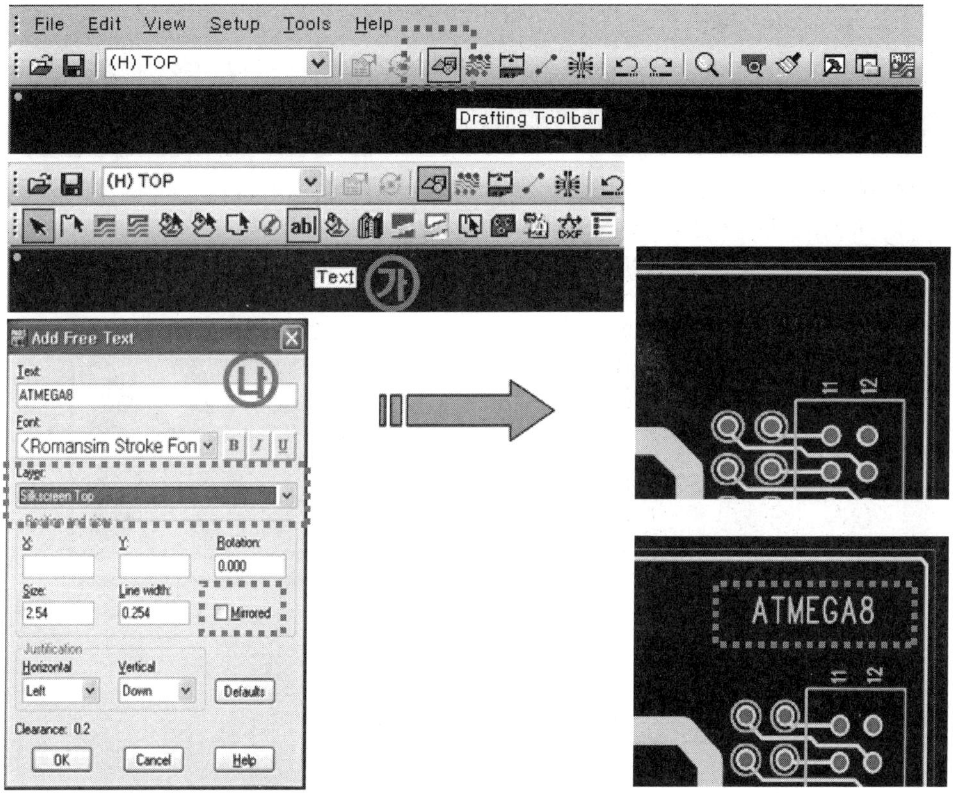

Text Mode

가. Drafting Toolbar -Text

- PCB에 보드네임이나 글자를 추가 할 수 있다.

나. Drafting Toolbar -Text 선택 -Add Free Text -Text (보드네임) -Layer(생성층)*Silkscreen TOP* – Size (Text 크기) -Line width (Text 두께) – OK

- Text 생성시 글자가 보이지 않을 시: Silkscreen TOP의 Text 색깔 변경 필수
- 단, BOT에 Text 추가 시 Layer (Silkscreen Bottom), Mirrored 체크 필수

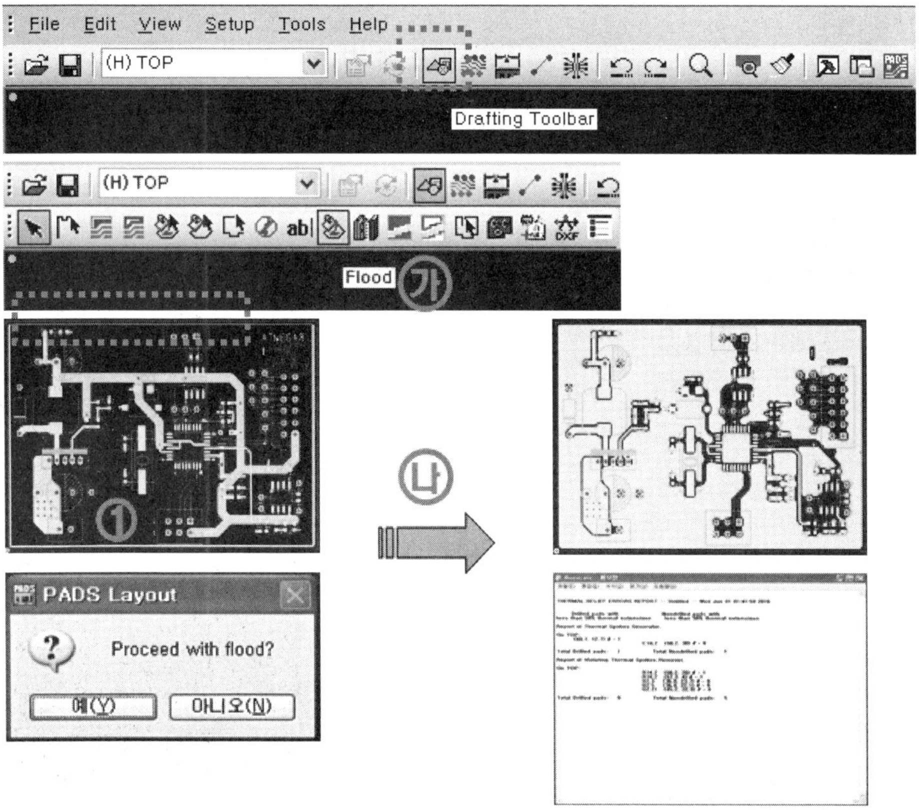

Flood Mode

가. Drafting Toolbar -Flood

- 빈 공간에 Copper pour을 채울 수 있다.

나. Drafting Toolbar -Flood선택 -Copper Pour 외각선 선택-PADS Layout -예 -Therm. Err - 닫기

- Copper pour의 외각라인을 따라 빈 공간에 Copper가 채워진다.
- 〈PO〉+Enter key : ①의 그림처럼 다시 돌아와 외각라인의 수정이 가능하다.

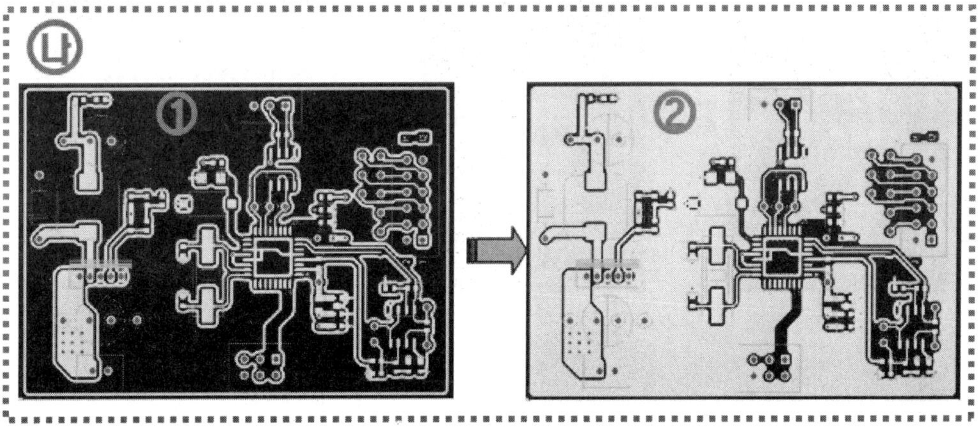

Hatch Mode

가. Drafting Toolbar – Hatch

- Copper Pour의 외각 라인을 채울 수 있다.

나. Drafting Toolbar – Hatch – Copper Pour 외각선 선택

- PCB파일을 Save한 후 재Open을 하면 ①의 그림처럼 Copper Pour의 외각 라인만 보여지게 된다.
- ②의 그림처럼 마지막 작업파일과 동일한 Copper Pour를 재현할 수 있다.

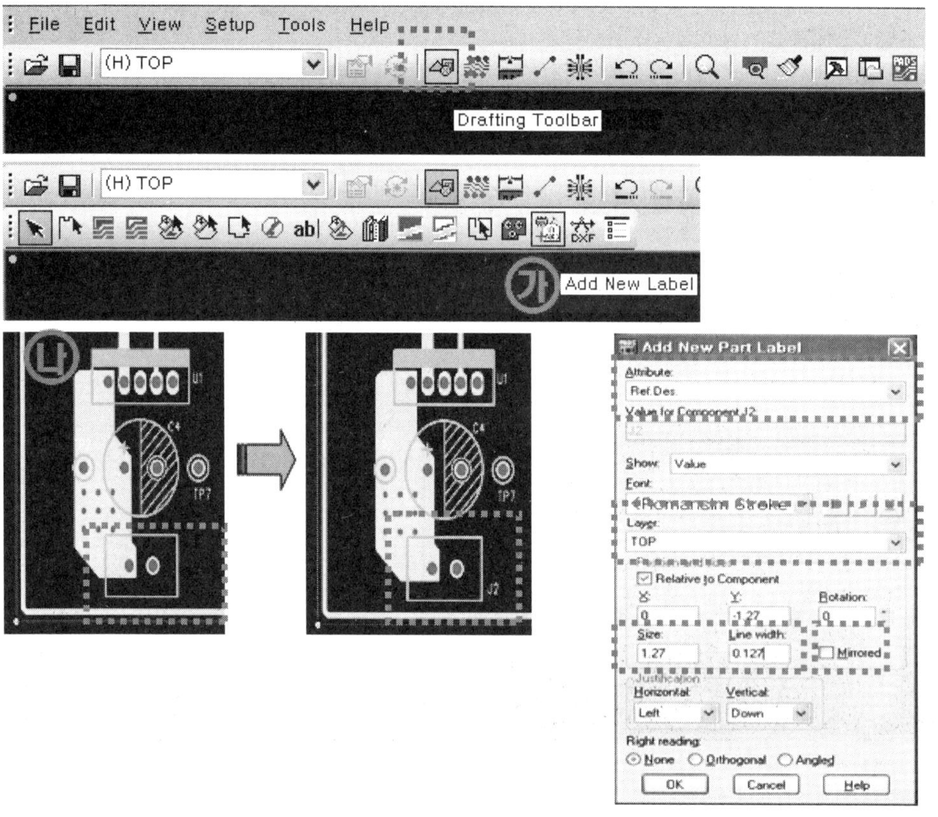

Add New Label Mode

가. Drafting Toolbar -Add New Label

• 삭제된 부품의 Ref. Des를 재생성 할 수 있다.

나. Drafting Toolbar -Add New Label 선택 -부품선택 -Add New Part Label -Attribute (Ref. Des=부품번호) -Layer (생성 층) – Size (Label크기) -Line width(Label두께) -OK -Label위치 변경

• 단. BOT에 Label 추가 시 Layer (Bottom). Mirrored 체크 필수

Layout 이해(Design Toolbar)

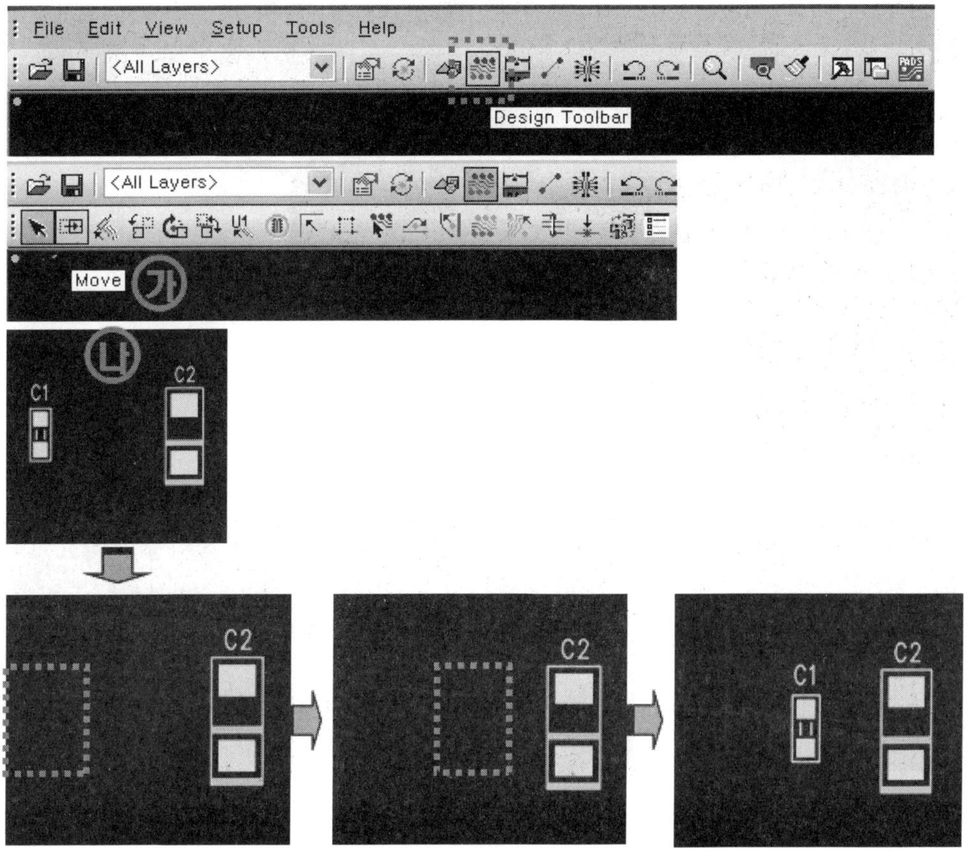

Move Mode

가. Design Toolbar -Move

- 부품을 원하는 좌표에 이동시킬 수 있다.

나. Design Toolbar -Move 선택 -부품선택 -이동 (Grid의 간격으로 이동)

- Grid의 간격으로 움직이므로 Grid값을 수정하여 이동한다.

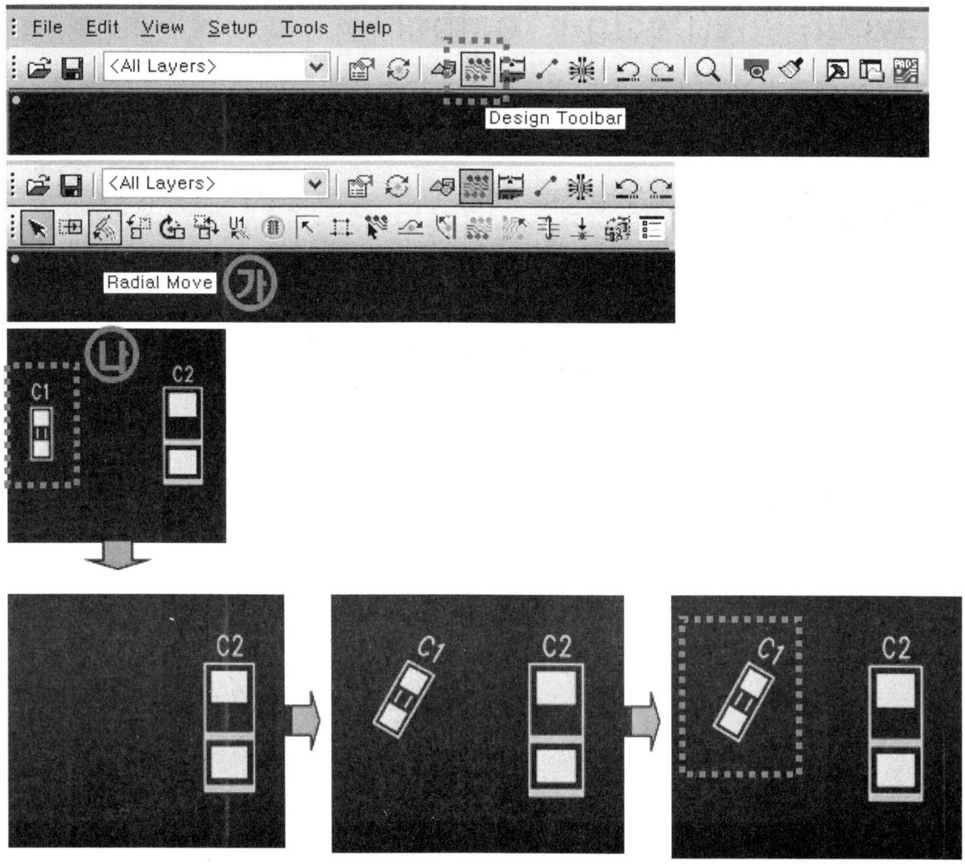

Radial Move Mode

가. Design Toolbar -Radial Move

- 부품을 방사형 좌표에 이동시킬 수 있다.

나. Design Toolbar -Radial Move 선택 -부품선택 -이동 (방사형으로 이동)

- 방사형 부품 배치에 활용이 가능하다.
- 부품의 각도는 방사형의 위치에 따라 달라진다.

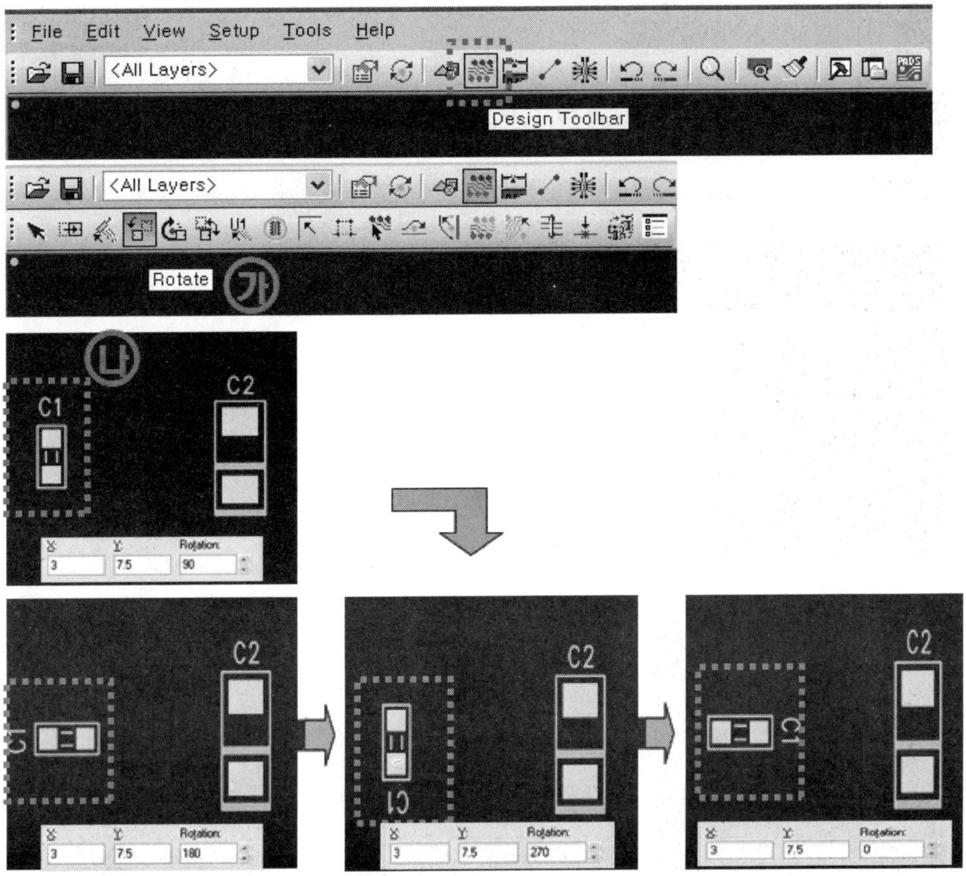

Rotate Mode

가. Design Toolbar -Rotate

- 부품을 각도를 변경할 수 있다. (0°,90°,180°,270°)

나. Design Toolbar -Rotate 선택 -부품선택 -각도변경

- 부품 배치에 활용이 가능하다.

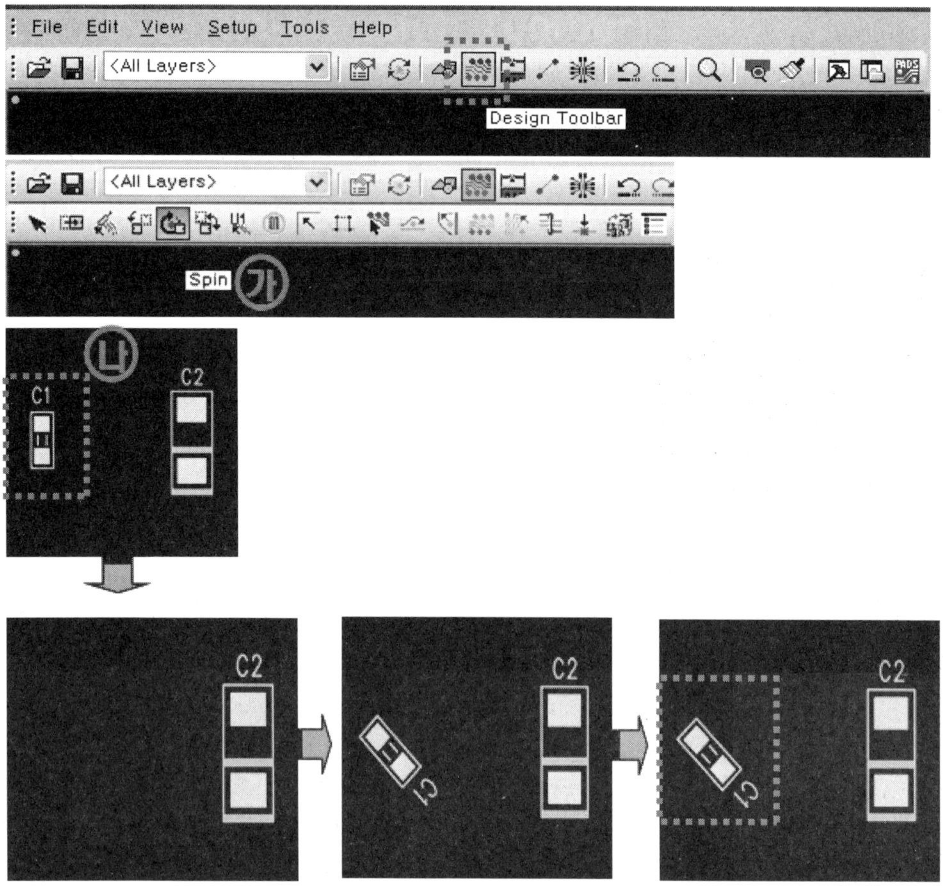

Spin Mode

가. Design Toolbar -Spin

- 부품을 각도를 변경할 수 있다. (0°~359°)

나. Design Toolbar -Spin **선택** -**부품선택** -**각도변경**

- 각도를 자유롭게 변경 가능하여 부품 배치에 활용이 용이하다.

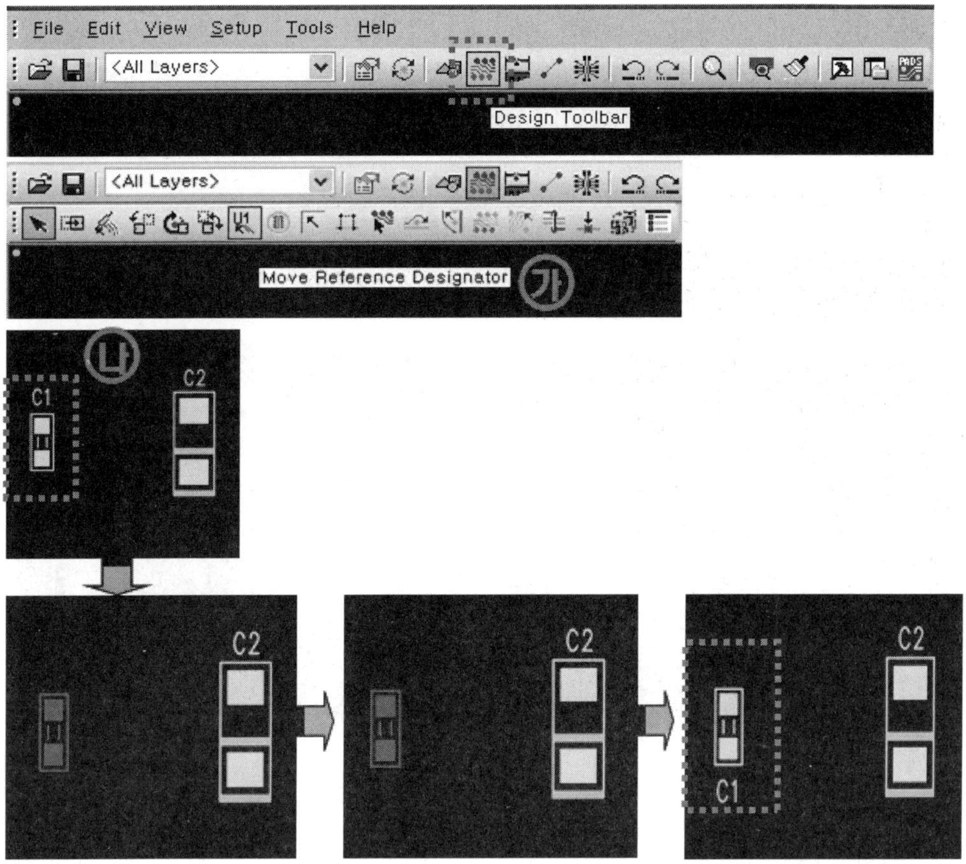

Move Reference Designator Mode

가. Design Toolbar -Move Reference Designator

- 부품을 Ref.Des.의 위치를 변경할 수 있다.

나. Design Toolbar -Move Reference Designator 선택 -부품 Ref.Des.선택 -이동

- 부품번호(=Ref.Des.)만 선택하여 이동이 가능하다.

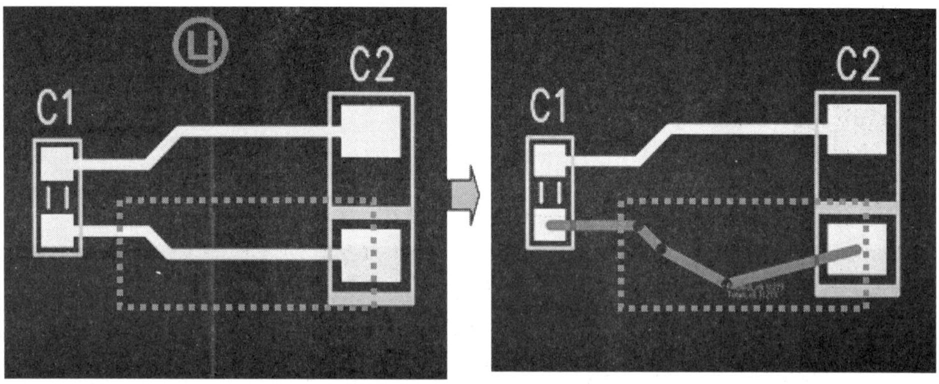

Add Corner Mode

가. Design Toolbar -Add Corner

- 패턴의 수정 시 사용할 수 있다.

나. Design Toolbar -Add Corner 선택 -수정할 패턴 선택

- 패턴의 각도를 변경하여 수정할 수 있다.

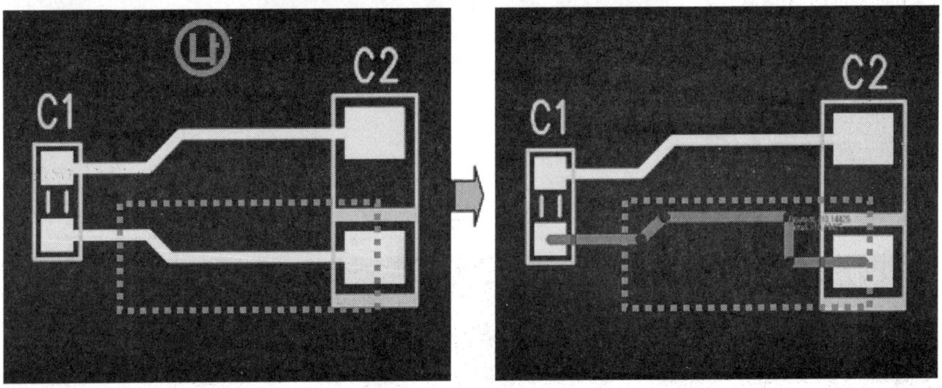

Split Mode

가. Design Toolbar -Split

- 패턴의 수정 시 사용할 수 있다.

나. Design Toolbar -Split **선택 -수정할 패턴 선택**

- 패턴의 각도를 변경하여 수정할 수 있다.

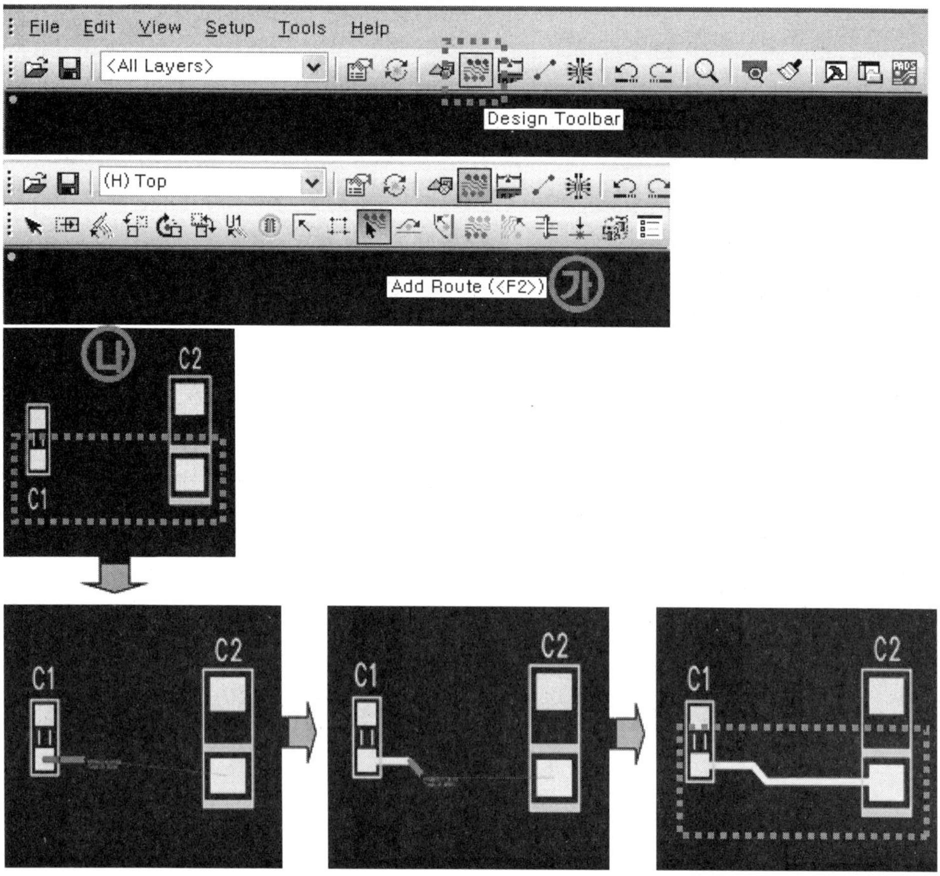

Add Route Mode

가. Design Toolbar -Add Route

- Connection으로 연결된 Net를 패턴화 시킬 수 있다.

나. Design Toolbar -Add Route 선택 -시작 Pin 선택 -끝맺음 Pin 선택

- 패턴의 각도를 변경하려면 마우스(좌측)을 클릭하여 원하는 각도로 연결한다.

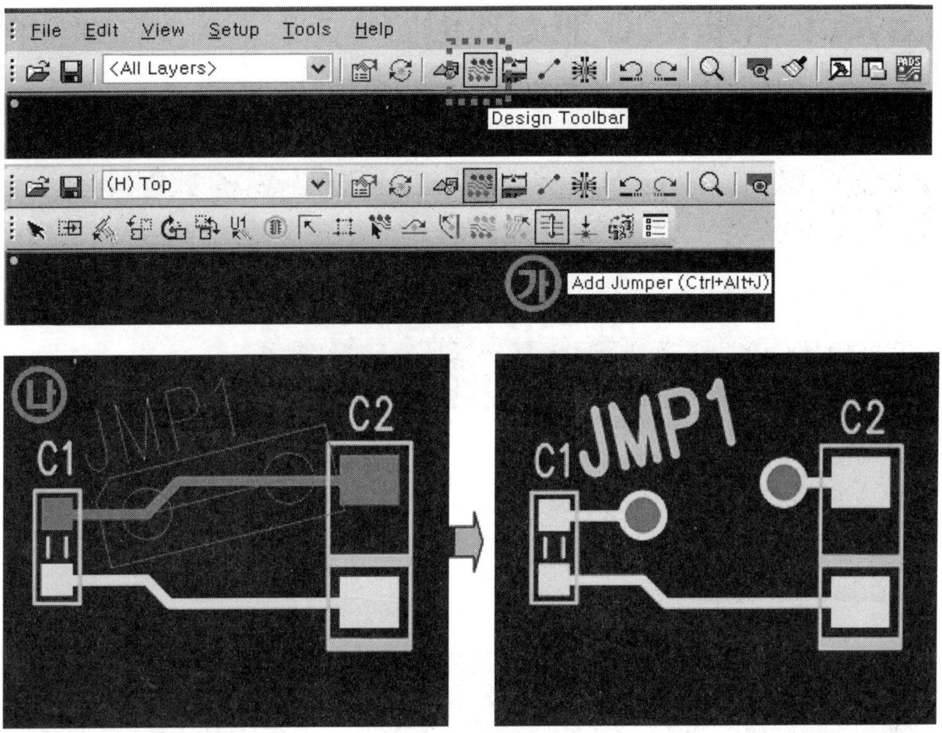

Add Jumper Mode

가. Design Toolbar -Add Jumper

- Jumper를 추가할 수 있다.

나. Design Toolbar -Add Jumper 선택 -수정할 패턴 선택

- 패턴에 Jumper를 연결할 수 있다.

Layout 이해(ECO Toolbar)

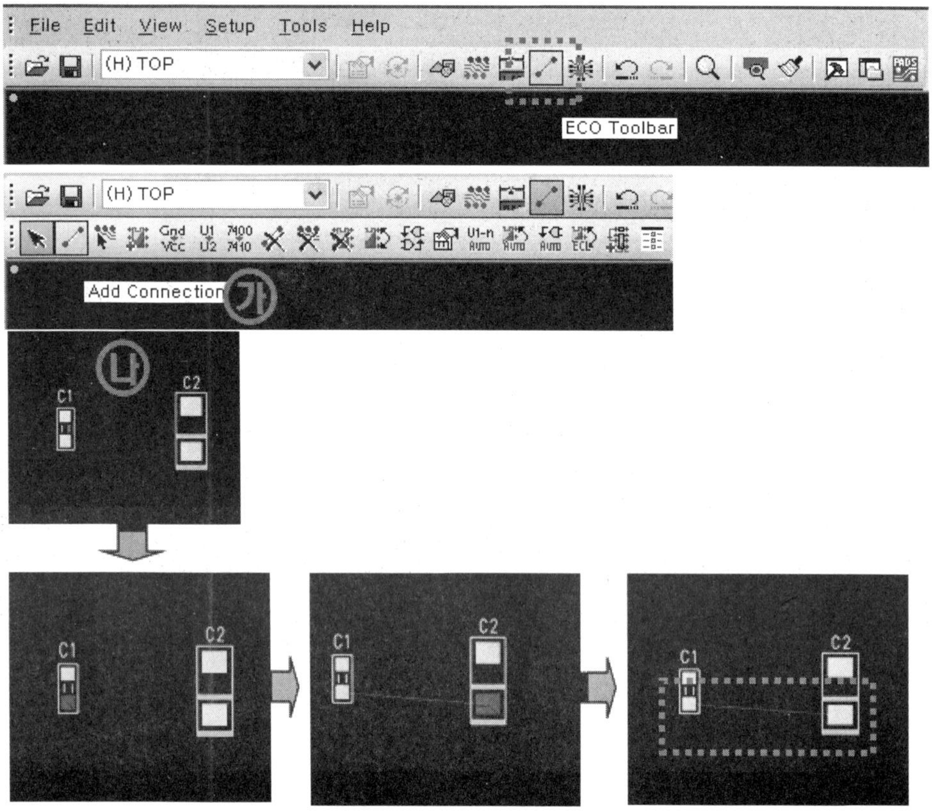

Add Connection Mode

가. ECO Toolbar -Add Connection

- 부품을 Connection으로 연결하여 네트를 형성 할 수 있다.

나. ECO Toolbar -Add Connection -시작 Pin선택 - 끝맺음 Pin 선택 – 〈ESC〉

- ECO 모드로 네트를 추가할 시 네트네임은 $$$~으로 시작한다.
- 회로와 매칭시키기 위해서는 추가된 Net name의 수정이 필요하다.
- Net name이 서로 다른 네트를 연결하려면 2개의 name중 하나를 선택해야 한다.

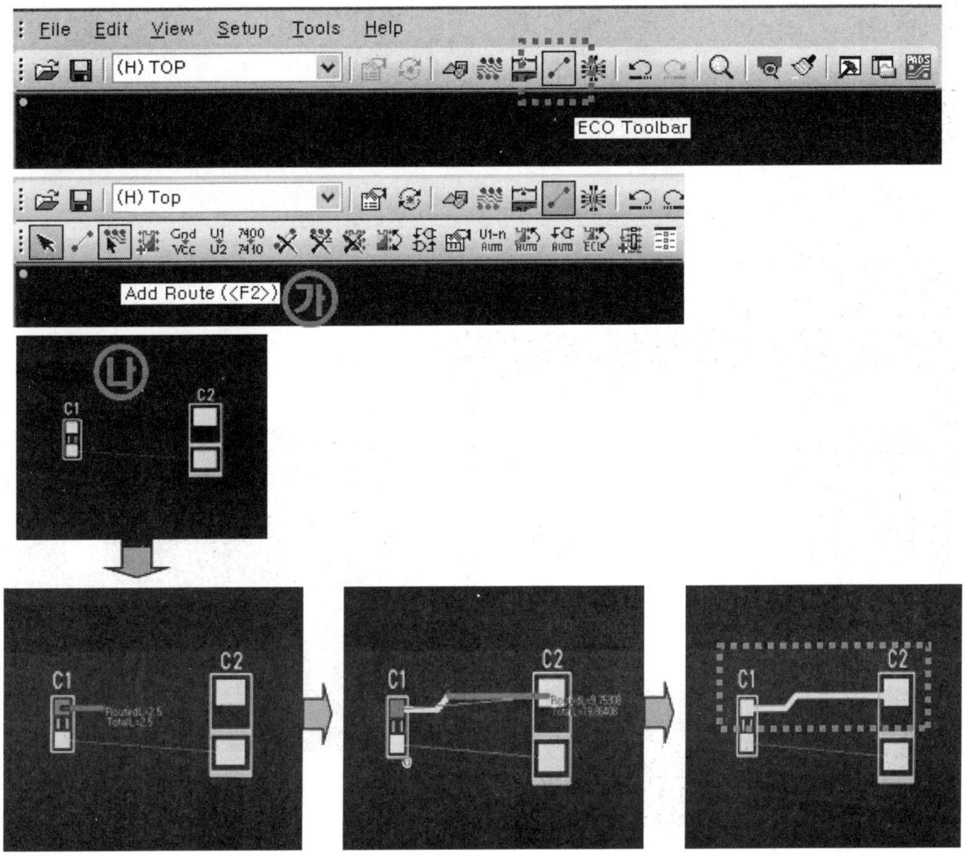

Add Route Mode

가. ECO Toolbar -Add Route

- Connection으로 연결되지 않은 Net를 패턴으로 만들 수 있다.

나. ECO Toolbar -Add Route -시작 Pin선택 - 끝맺음 Pin 선택

- 일대일 연결된 회로 구성을 할 때 유용하게 사용 할 수 있다.
- Connection으로 연결하지 않은 net가 생성될 수 있어 주의가 필요하다.

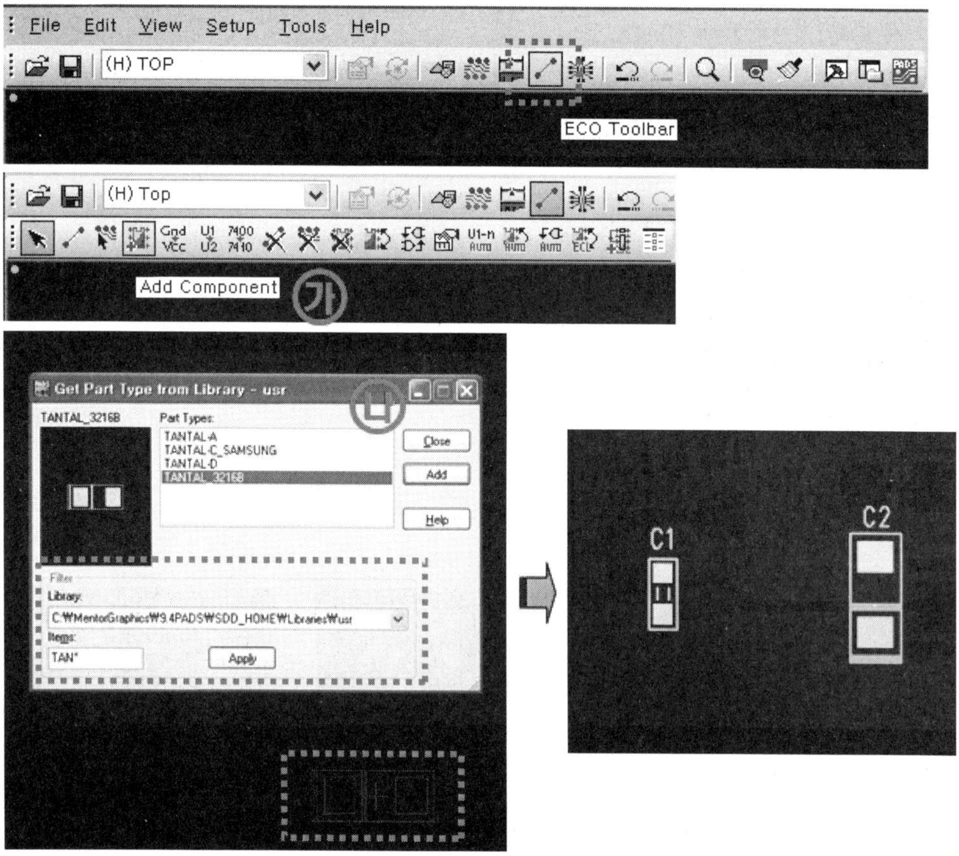

Add Component Mode

가. ECO Toolbar -Add Component

- 완성된 회로에 부품을 추가할 수 있다.

나. ECO Toolbar -Add Component -Get Part Type from Library – Library (부품 위치) -Items (부품 품명*) -Apply -Part Types: 선택 – Add

- 부품은 Library에 Part Type으로 저장이 되어 있을 때 추가가 가능하다.

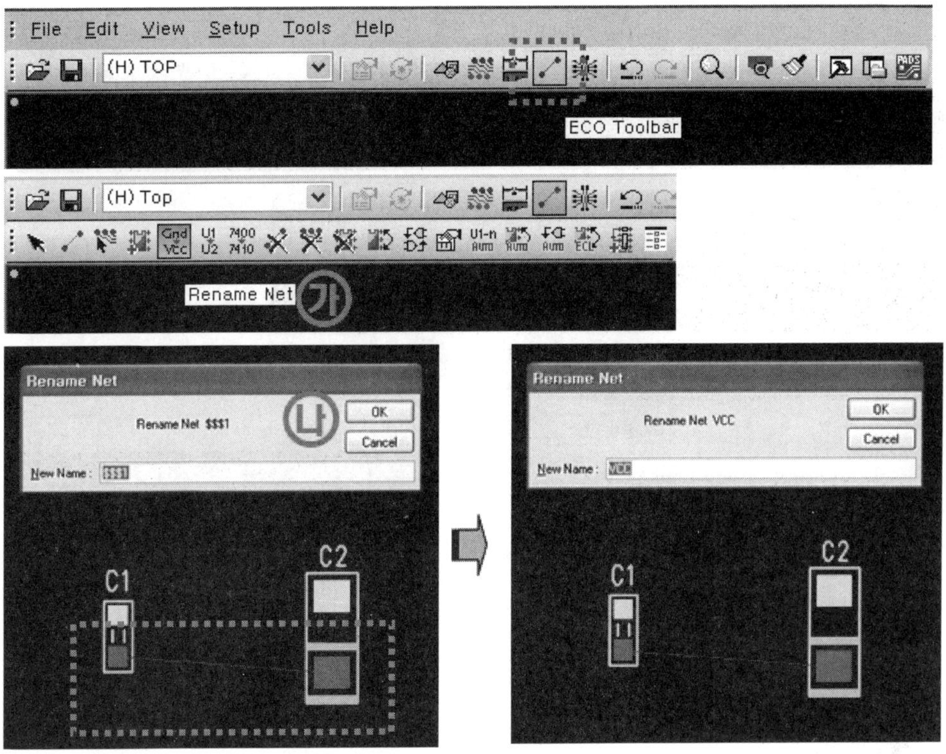

Rename Net Mode

가. ECO Toolbar -Rename Net

- 회로도의 변경 없이 Net Name을 수정할 수 있다.

나. ECO Toolbar -Rename Net 선택 -변경될 Net 선택 -Rename Net -New Name (기존Net->신규 Net name으로 변경) -OK

- 기존 회로에 동일한 Net Name이 있을 경우 변경되지 않는다.

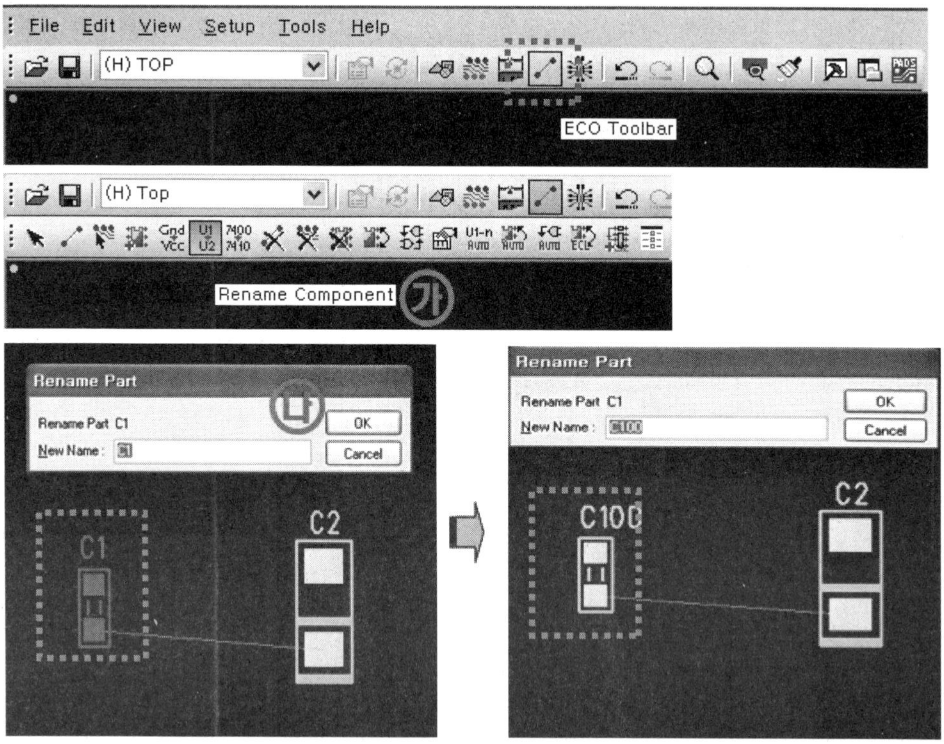

Rename Component Mode

가. ECO Toolbar -Rename Component

- 회로도의 변경 없이 부품 Ref.Des.를 수정할 수 있다.

나. ECO Toolbar -Rename Component 선택 -변경될 부품 선택 -Rename Part – New Name(기존Ref.->신규 Ref.으로 변경) -OK

- 기존 회로에 동일한 Ref.Des.이 있을 경우 변경되지 않는다.

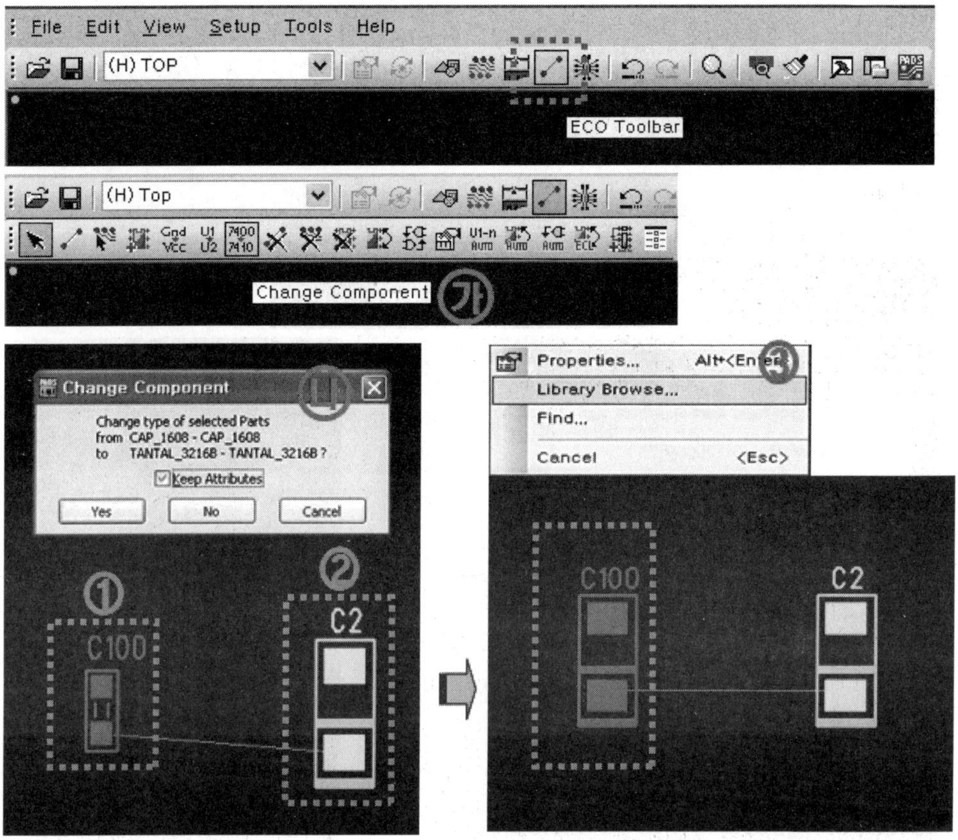

Change Component Mode

가. ECO Toolbar -Change Component

- 회로도의 변경 없이 부품을 변경할 수 있다.

나. ECO Toolbar -Change Component 선택 -① 변경될 부품 선택 -② 신규부품 선택 - Change Component (① 기존부품-> ② 신규 부품으로 변경 확인) - Yes

- ③ 작업파일에 존재하지 않는 부품일 경우 Library에서 검색하여 변경할 수 있다.

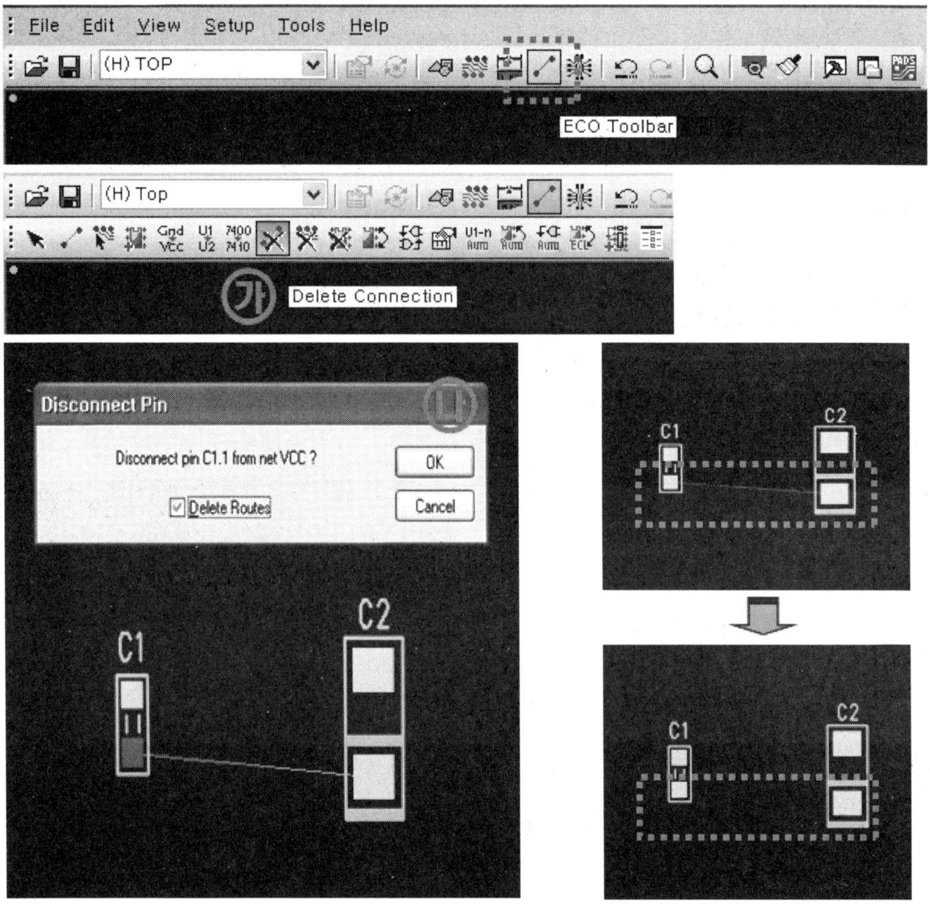

Delete Connection Mode

가. ECO Toolbar -Delete Connection

- 부품을 Connection으로 연결한 네트를 삭제 할 수 있다.

나. ECO Toolbar -Delete Connection -**삭제할** Pin **선택** -Disconnect Pin – OK

- 전체 Net가 삭제되는 경우도 있어 주의가 필요하다.

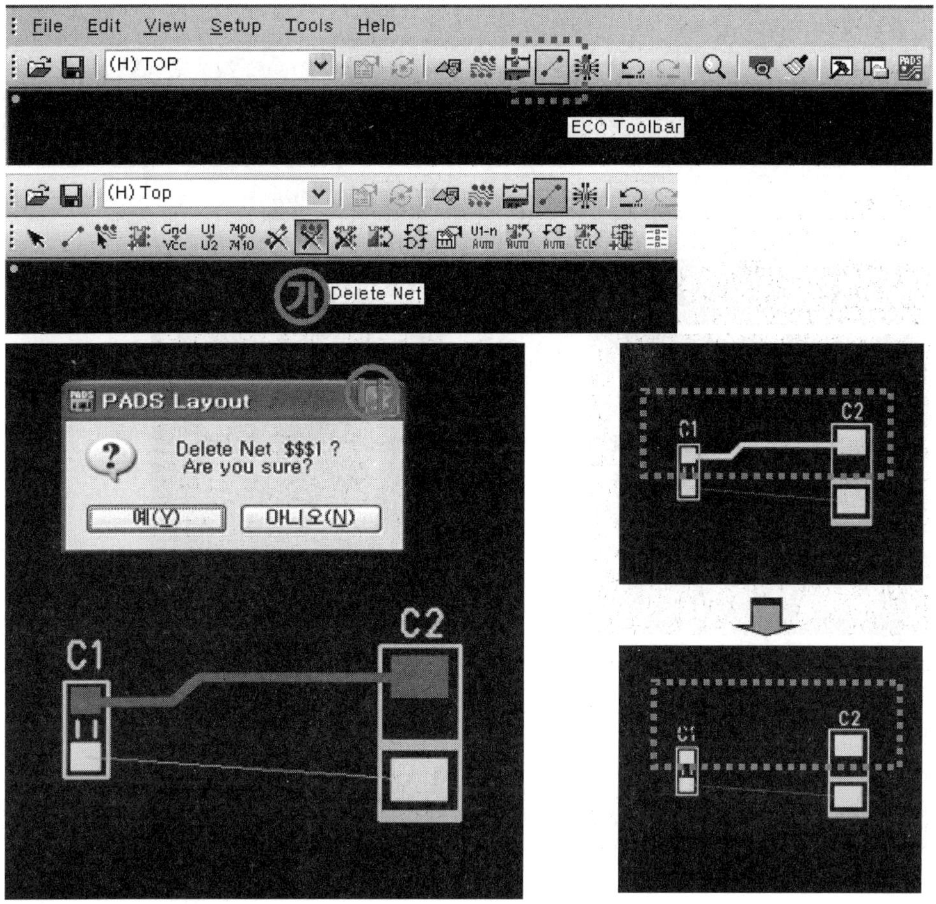

Delete Net Mode

가. ECO Toolbar -Delete Net

- 선택한 Net를 삭제할 수 있다.

나. ECO Toolbar -Delete Net -**삭제할** Net **선택** – PADS Layout (Net **확인**) – 예

- 특정한 Net의 전체를 삭제할 때 유용하게 사용할 수 있다.

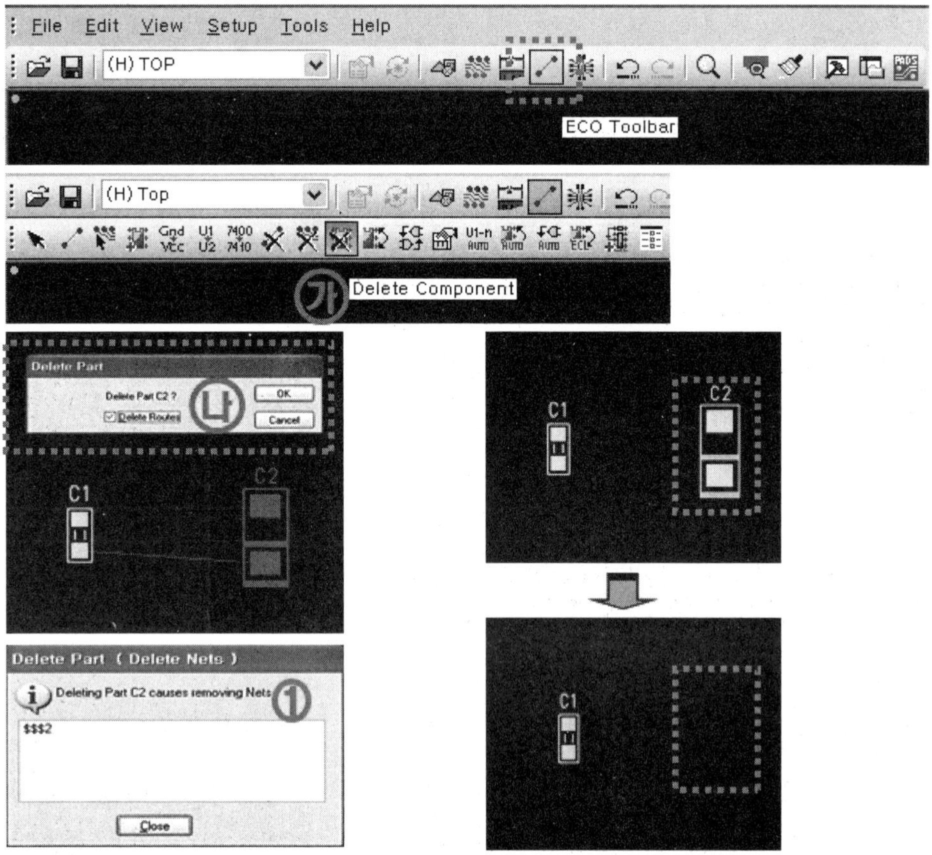

Delete Component Mode

가. ECO Toolbar -Delete Component

- 완성된 회로에 부품을 삭제 할 수 있다.

나. ECO Toolbar -Delete Component -삭제할 부품 선택 – Delete Part (삭제 부품 확인) -① Delete Part (Delete Nets) -Close

- ①번 그림과 같이 특정한 Net의 전체가 삭제될 때 확인이 필요하다.

07 / Layout 실습 (Net list)

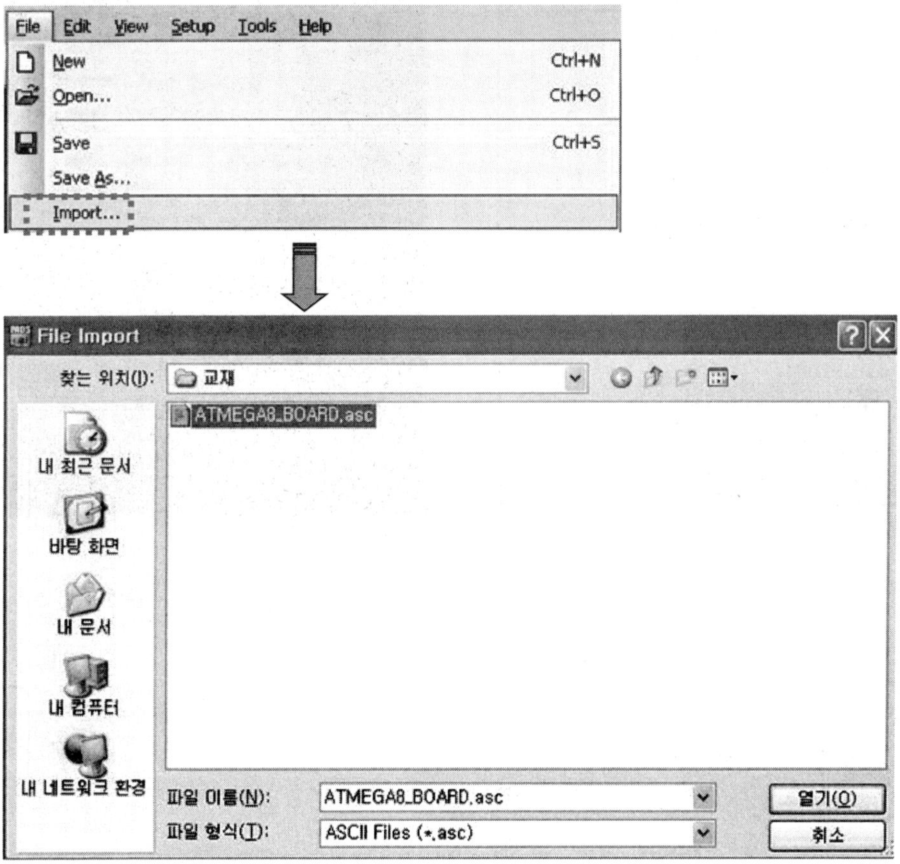

Net list Import

가. PADS Layout Icon – start a new design – File – Import – XXX.asc 열기

- 생성된 XXX.asc는 회로도와 동일한 경로이고 New에서 Import해야 한다.
- DATA 정보가 있는 File에서 Import시 기존 DATA와 겹치게 된다.

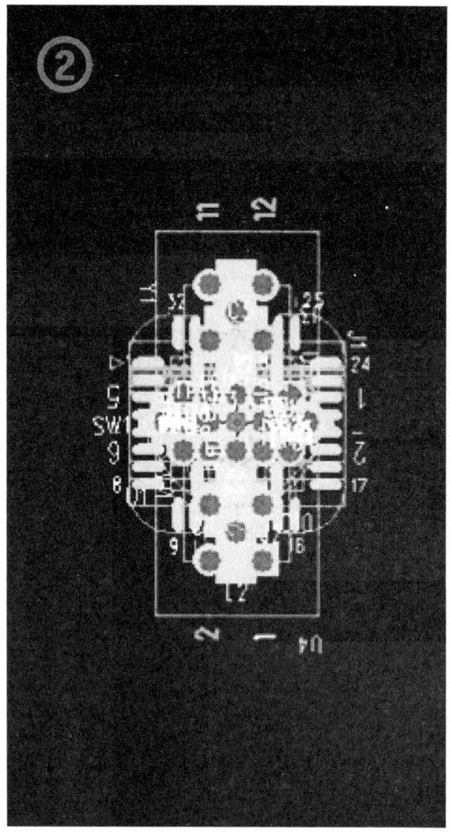

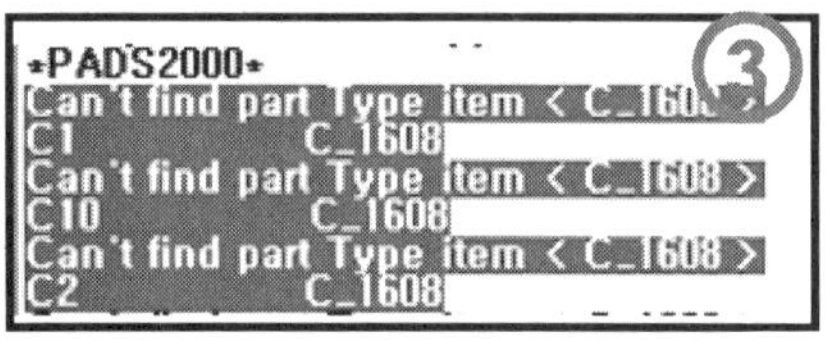

Net list Import

가. Import – XXX.asc 열기

- 생성된 XXX.asc는 ①의 그림과 같이 Error가 없어야 한다.
- ②과 같이 원점에 모든 부품이 겹쳐 있다.
- ③의 Error는 Library에 Part Type이 없거나 Footprint를 잘못 입력한 것이다.

Layout 실습 (Disperse)

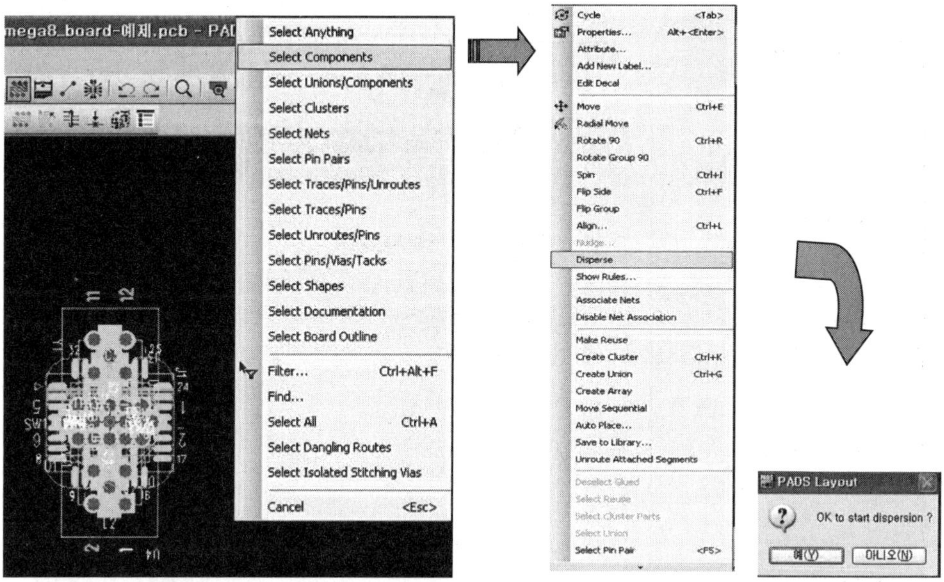

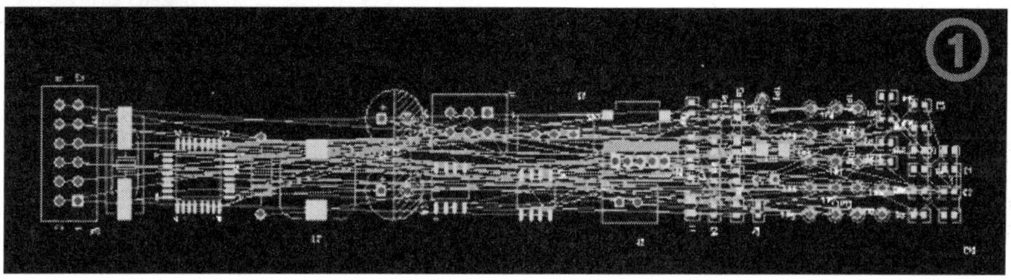

Layout (Disperse)

가. No Select - 마우스 (우측) – Select Components – 전체 부품 - 마우스 (우측) – Disperse – Ok to start dispersion – YES

- ①의 그림과 같이 부품이 나열된다.
- ①의 그림을 보고 참고하여 PCB 외형 Size를 유추할 수 있다.

Layout 실습(Unit 설정)

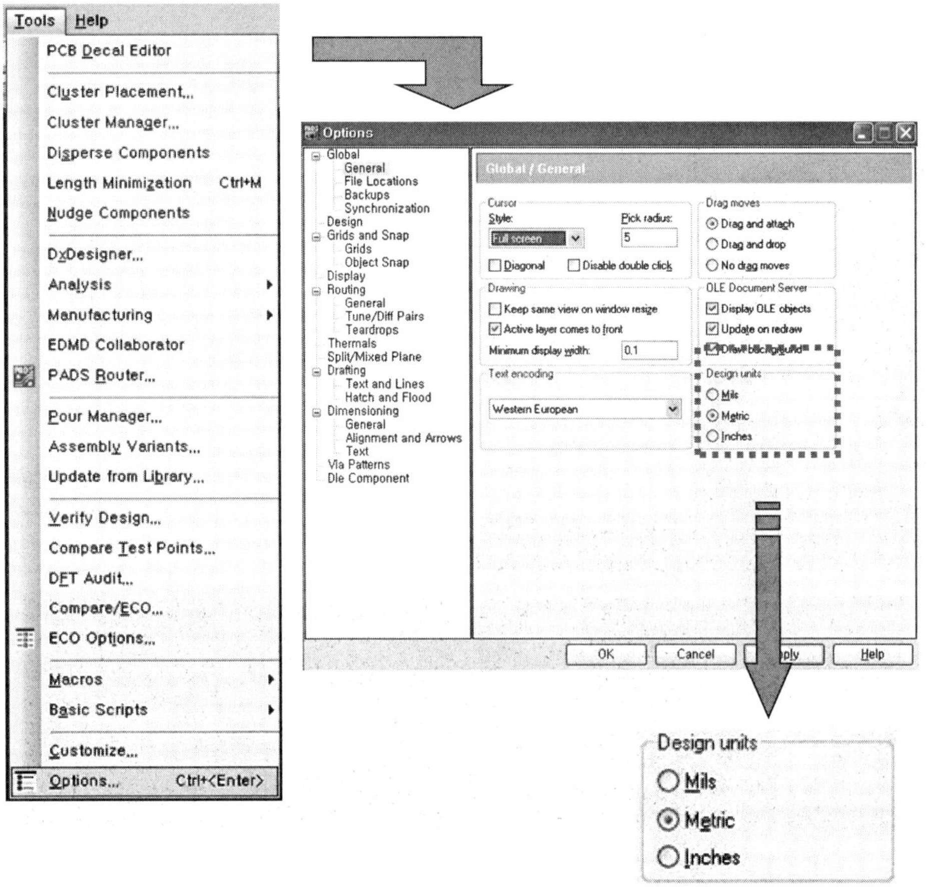

Layout (단위 설정하기)

가. Tools – Option – Global – General – Design units 선택

- Decal 작성시 단위 설정을 한다. (Mils, Metric (mm), Inches)

Layout 실습(외형 생성)

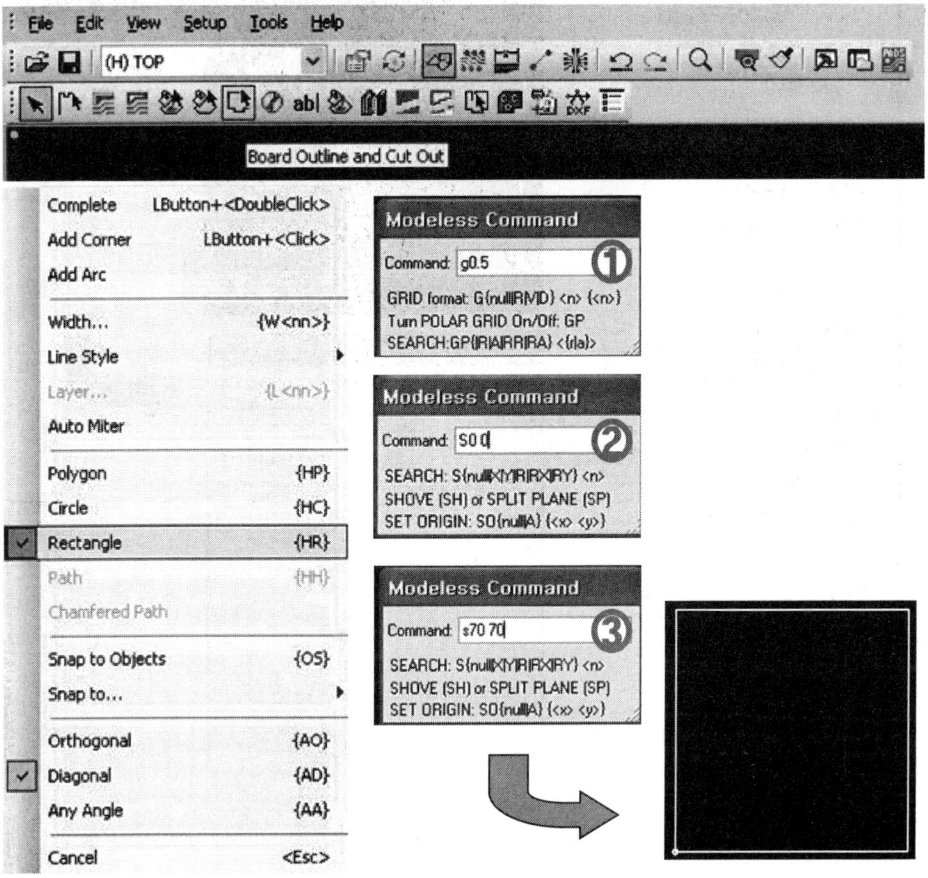

Board Outline 생성

가. Drafting Toolbar – Board Outline and Cut Out 선택 – 마우스(우측) – Rectangle – ① g0.5 (Grid 변경) – Enter – ② s0 0 (원점 이동) – Enter – Space – ③ s70 70 (외형 size) – Enter – Space

- 정확한 치수를 그리려면 Grid를 최소단위로 설정 후에 작업해야 한다.

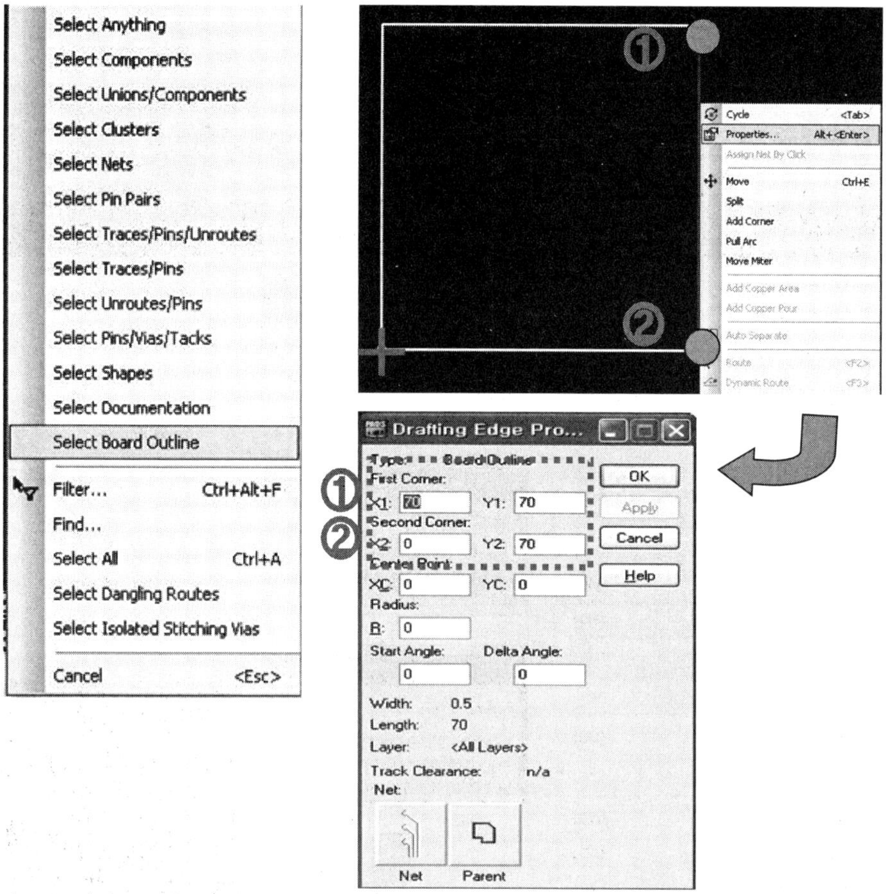

Board Outline 생성

가. 마우스(우측) – Select Board Outline 선택 – 외형 1면 – Properties – 외형 Size 확인 및 수정 – OK

- 원점을 기준으로 선택한 선의 ① 위 / ② 아래 Point의 좌표를 변경한다.

Layout 실습 (부품 배치)

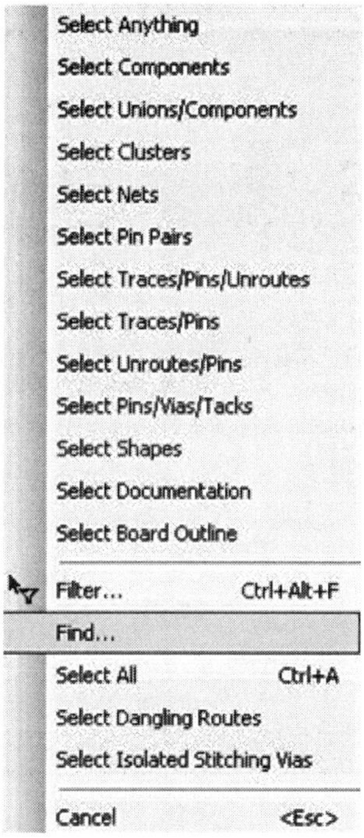

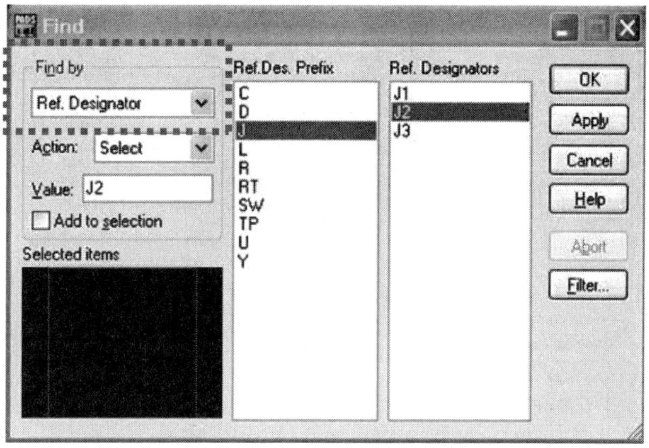

Layout (배치)

가. 마우스(우측) – Find 선택 – Ref. Des. Prefix(Ref의 알파벳 선택) – Ref. Designators – **배치 부품 선택**

- 부품의 배치는 기본적으로 회로를 기준으로 한다.
- 전체 배치를 하기에 앞서 부분 배치 → 전체 배치로 작업방향을 잡는다.
- Net별, PCB Decal, Part Type 별 선택이 가능하므로 배치에 활용한다.
- 회로의 흐름을 먼저 파악한 후에 배치를 하는 것이 효율적이다.
- 양면 배치일 경우 겹쳐서 배치를 하여 공간 효율을 높일 수 있다.

Layout 실습 (Glued 활용)

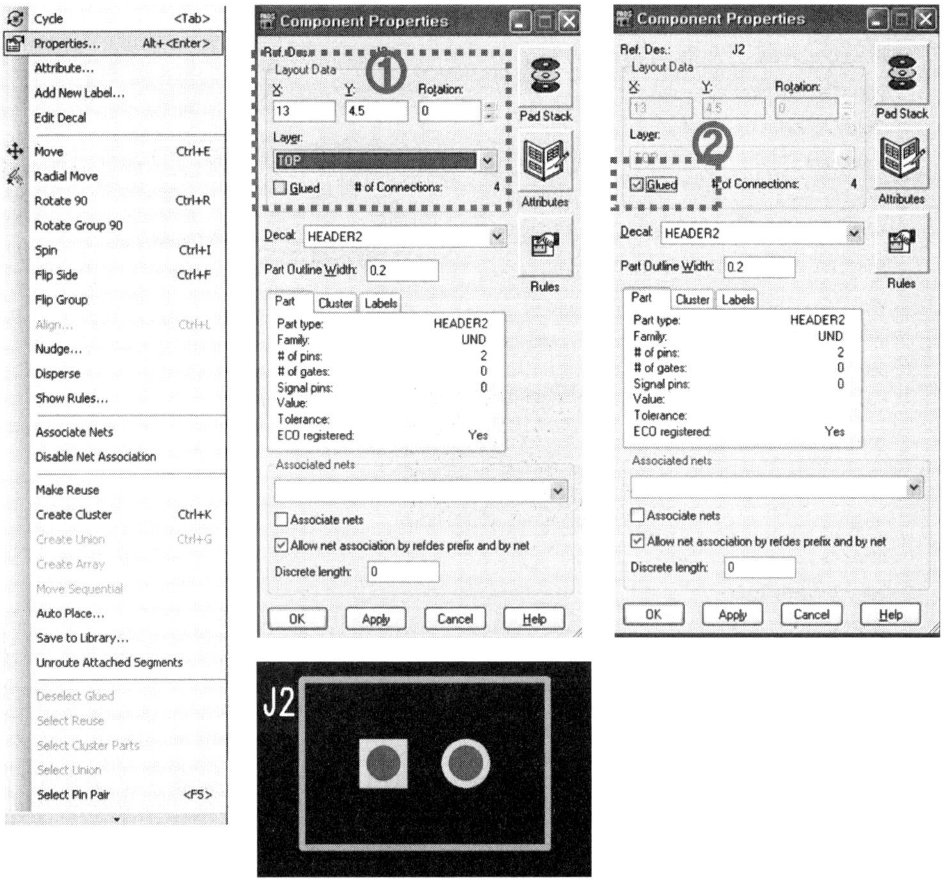

부품 Glued 활용

가. 마우스(우측) – Select Component 선택 – 부품 선택 – 마우스(우측) – Properties – ① 부품의 좌표 및 Layer 설정 – ② Glued 선택 – OK

- 기구적으로 특정위치가 정해지면 ②와 같이 부품을 고정시킨다.

Layout 실습 (Copper)

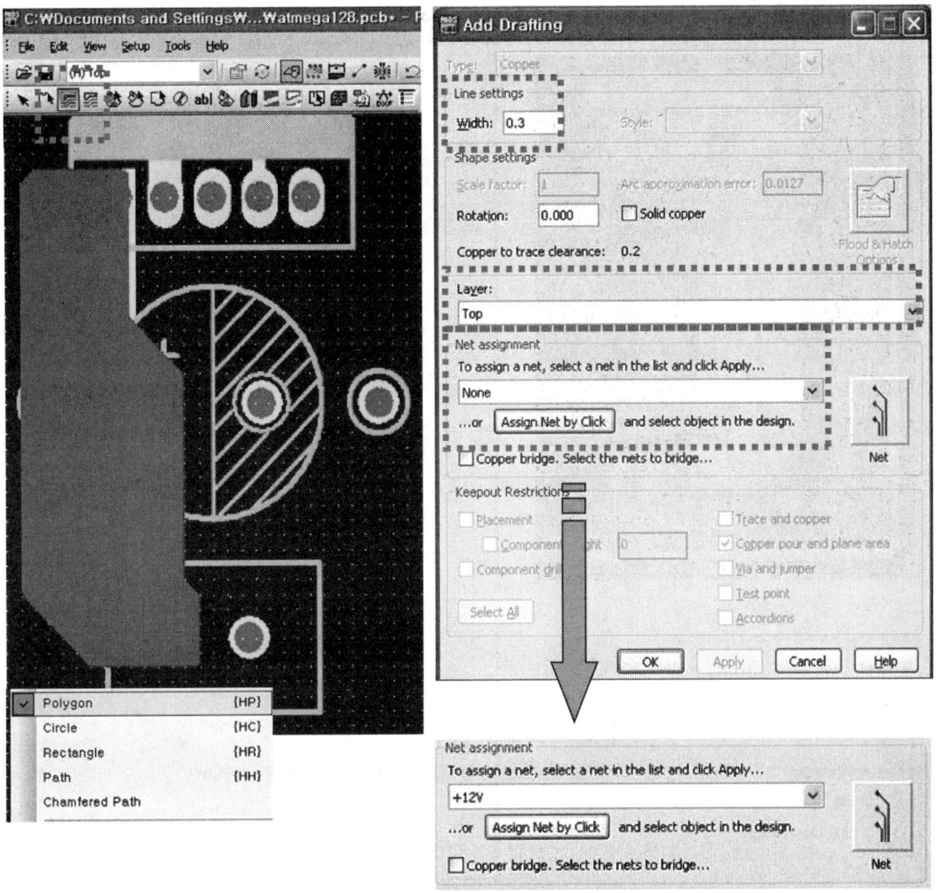

Copper 생성

가. Drafting Toolbar – Copper 선택 - 마우스(우측) – Polygon 선택 – 도형 생성 – Add Drafting (Width=0.3/Layer(Top)/Net assignment(12V 선택)) – OK

- Copper를 이용하여 두꺼운 전원 Line의 배선을 생성할 수 있다.

Layout 실습 (Copper pour)

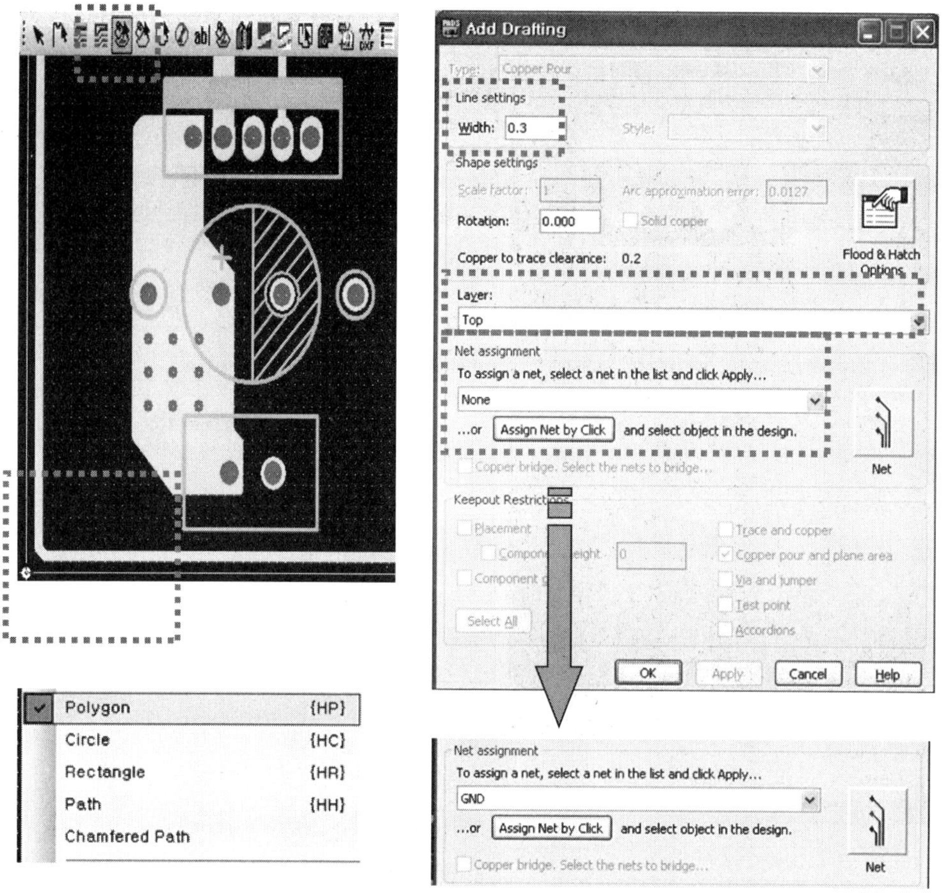

Copper Pour 생성

가. Drafting Toolbar – Copper Pour 선택 – 마우스(우측) – Polygon 선택 – 도형 생성 – Add Drafting (Width=0.3/Layer(Top)/Net assignment(GND 선택)) – OK

- Copper Pour를 이용하여 GND의 배선을 생성할 수 있다.

Layout 실습(Design Toolbar)

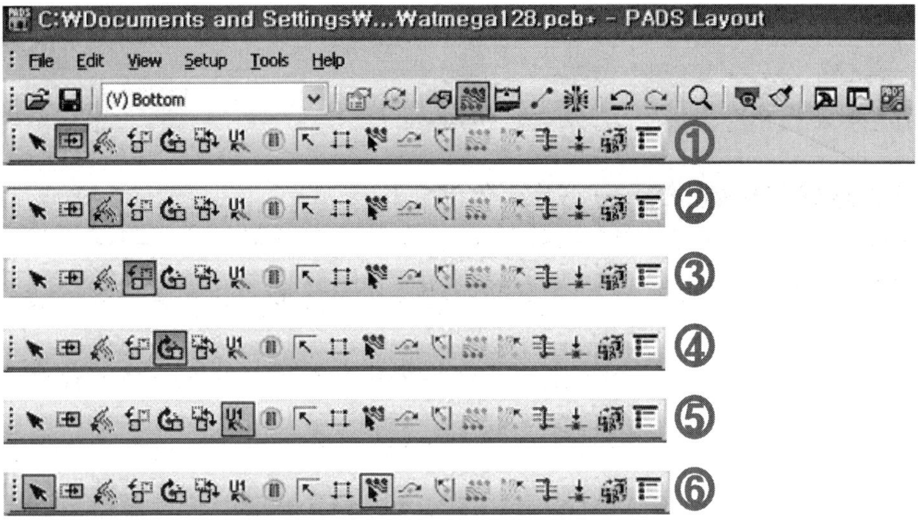

Design Toolbar

가. Design Toolbar

① Design Toolbar - Move(부품 위치 이동)

② Design Toolbar - Radial Move(부품 방사형 이동)

③ Design Toolbar - Rotate(부품 90° 방향전환)

④ Design Toolbar - Spin(부품 미세 각도 방향전환)

⑤ Design Toolbar - Move Reference Designator(부품 Ref. 이동)

⑥ Design Toolbar - Add Route(배선)

Layout 실습(ECO)

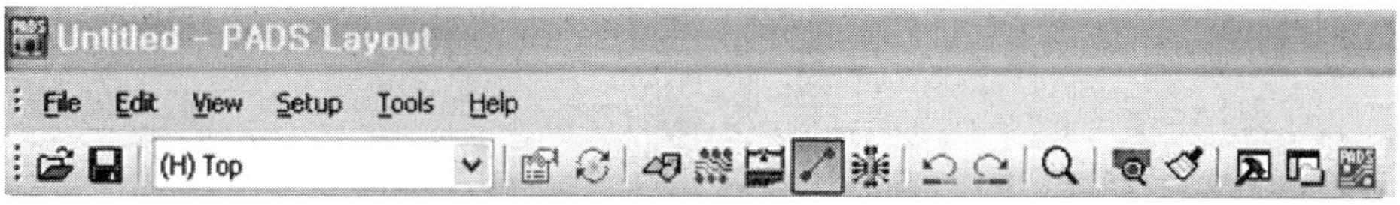

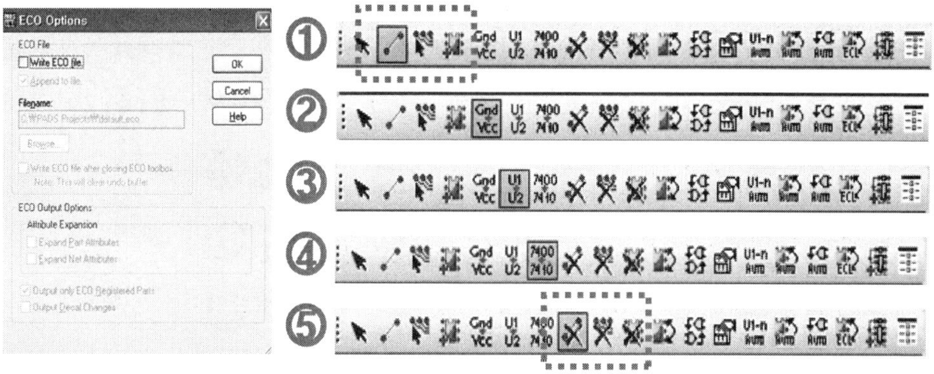

ECO Toolbar

가. ECO Toolbar

① ECO Toolbar – Add Connection/Route/Component (Item 추가)

② ECO Toolbar – Rename Net (Net Name 변경)

③ ECO Toolbar – Rename Component (Reference 변경)

④ ECO Toolbar – Change Component (부품 변경)

⑤ ECO Toolbar – Delete Connection/Route/Component (Item 삭제)

08 / Layout 실습 (CAM 이해) (1)

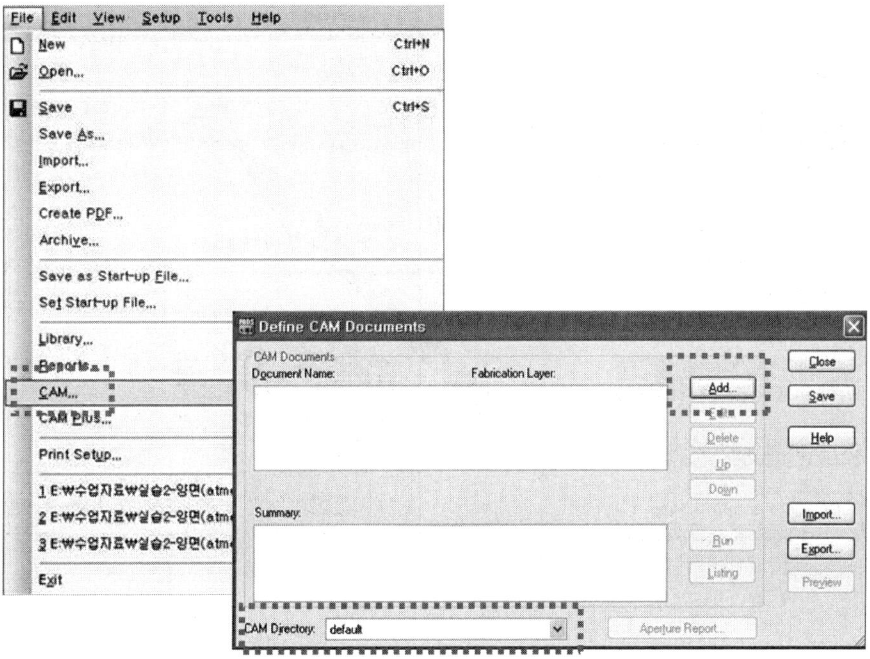

CAM (Gerber 생성)

가. File - CAM 선택 – Define CAM Documents – Add – Cam Directory (경로지정) – Save – Close

- CAM을 이용하여 Gerber를 생성한다.

나. 양면 PCB, 양면 배치일 경우 Gerber 수

1. Top (Pattern, Solder Mask, Silk screen)
2. Bottom (Pattern, Solder Mask, Silk screen)
3. Drill (Drill Drawing, NC Drill)

Layout 실습 (CAM ①)

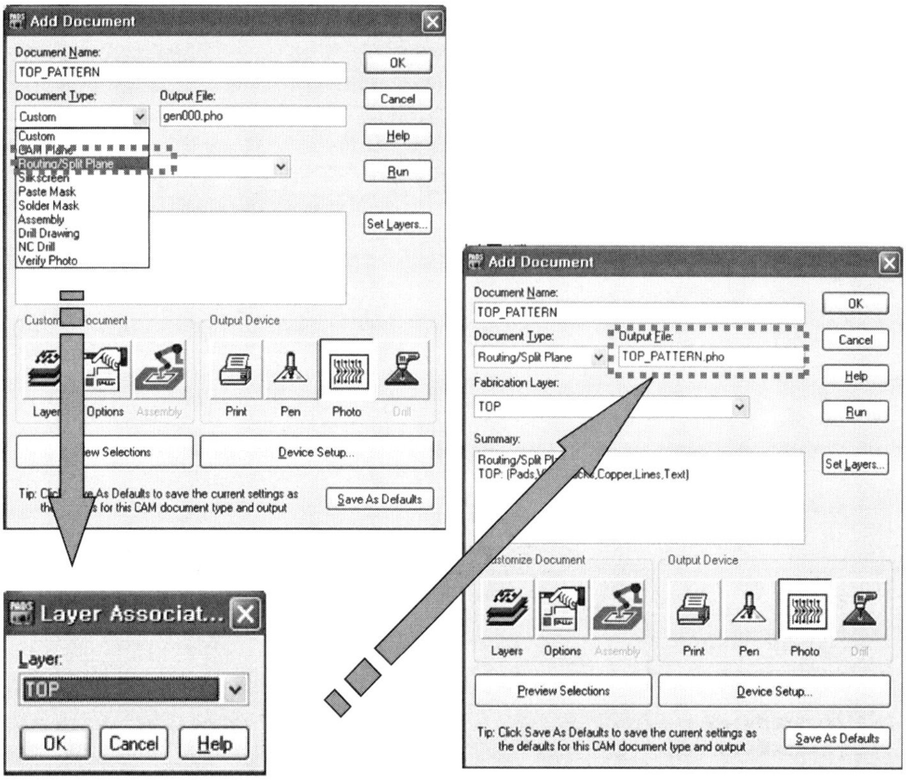

CAM (TOP_Pattern 생성)

가. Define CAM Documents – Add – Add Document – Document Name (TOP_PATTERN) – Document Type (Routing Split Plane선택) – Layer Associate – Layer (TOP 선택) – OK – Add Document – Output File (Document Name 동일)

- PCB 제작에 필요한 Gerber DATA는 Output File Name(.pho)이다.

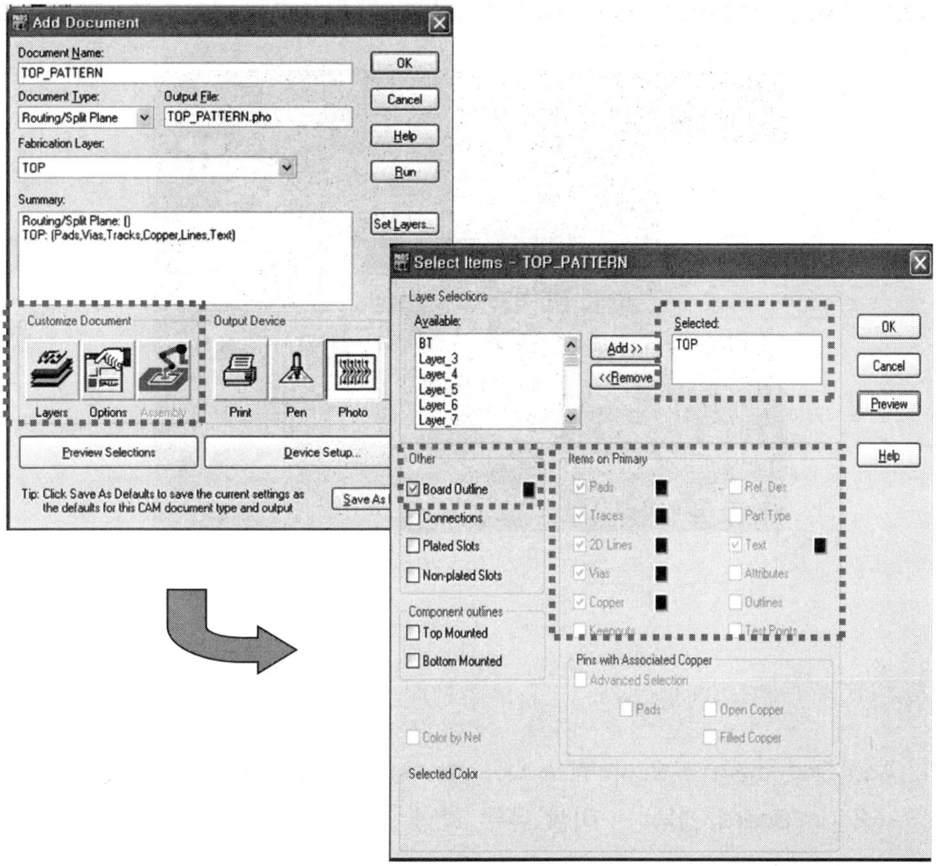

CAM (TOP_Pattern 생성)

가. Add Document – Customize Document (Layer 선택) – Other (Board Outline) – Selected (TOP 선택) – Item on Primary (Pads, traces, 2D Line, Vias, Copper, Text 선택 확인) – OK

- 각 Gerber에 해당하는 Item이 선택되어 있는지 확인한다.

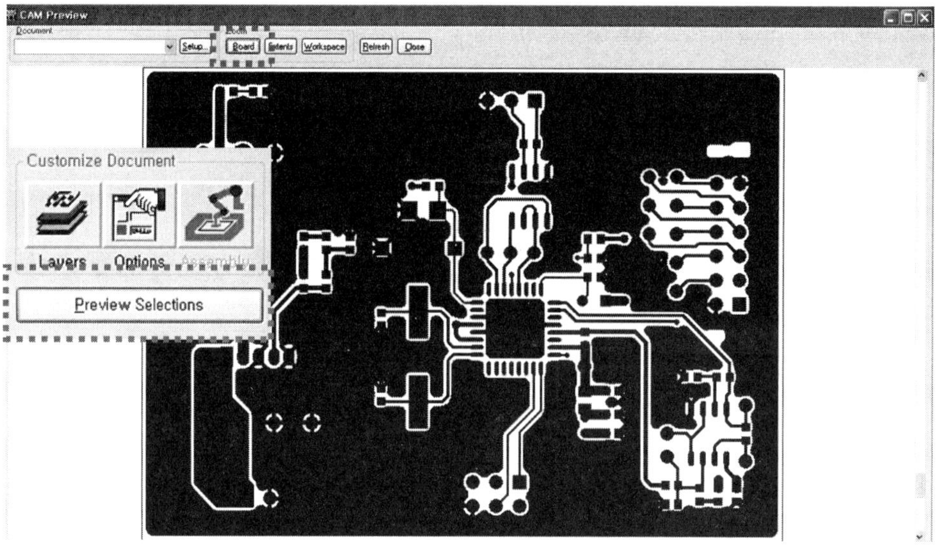

CAM (TOP_Pattern 생성)

가. Add Document – Customize Document (Preview Selections 선택) – Zoom(Board 선택) – 이상 유무 확인 – Close

- TOP Pattern의 Gerber를 확인 후에 저장한다.

Layout 실습 (CAM ②)

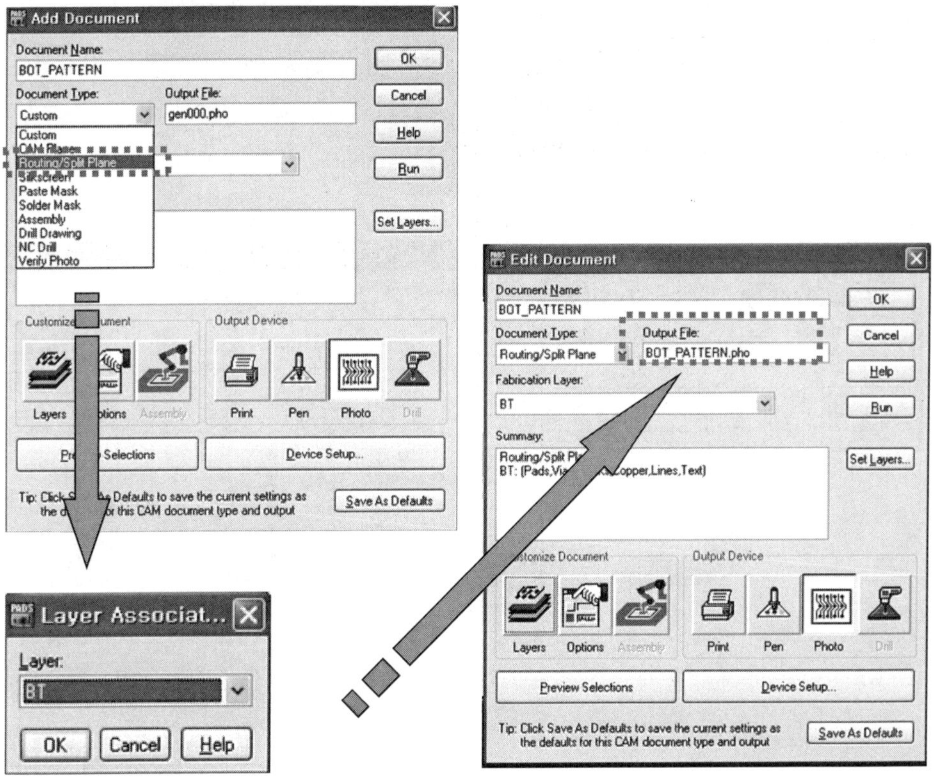

CAM (BOT_Pattern 생성)

가. Define CAM Documents – Add – <u>Add Document</u> – Document Name (BOT_PATTERN) – Document Type (Routing Split Plane 선택) – Layer Associate – Layer (BOT 선택) – OK – <u>Add Document</u> – Output File (Document Name 동일)

- PCB 제작에 필요한 Gerber DATA는 Output File Name(.pho)이다.

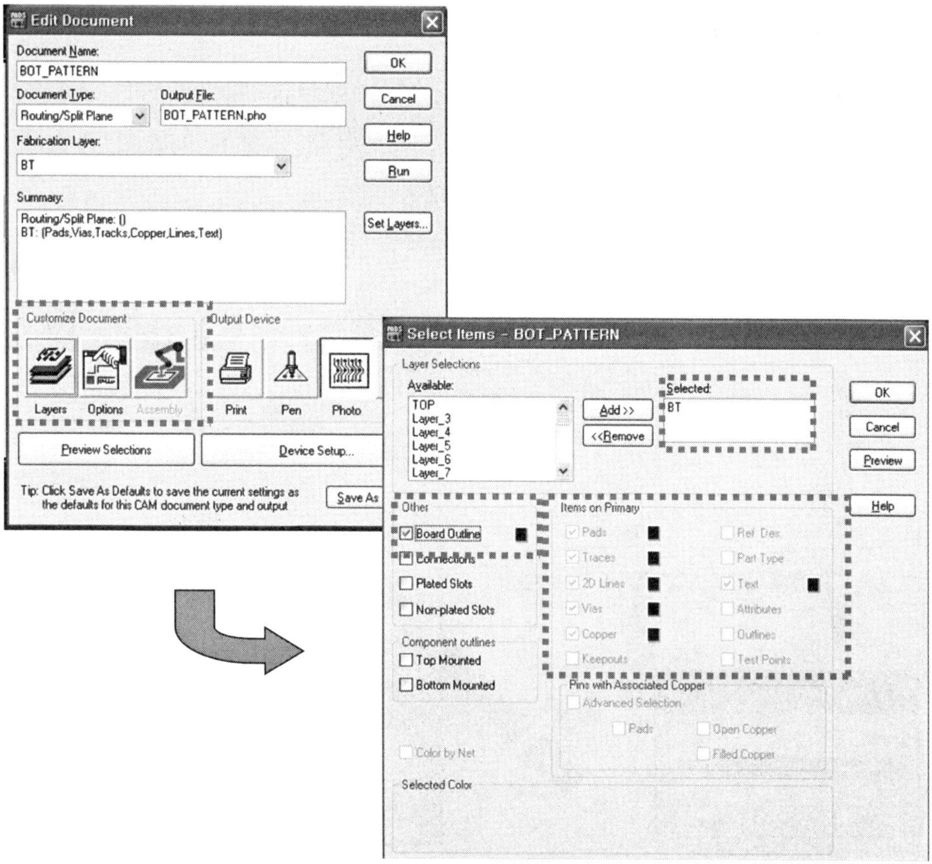

CAM (BOT_Pattern 생성)

가. Add Document – Customize Document (Layer 선택) – Other (Board Outline) – Selected (BOT 선택) – Item on Primary (Pads, traces, 2D Line, Vias, Copper, Text 선택 확인) – OK

- 각 Gerber에 해당하는 Item이 선택되어 있는지 확인한다.

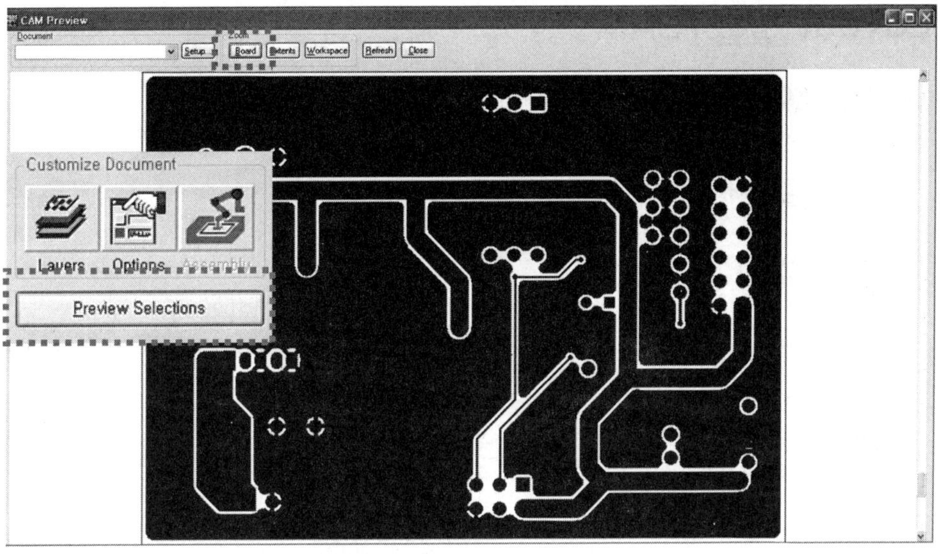

CAM (BOT_Pattern 생성)

가. Add Document – Customize Document (Preview Selections 선택) – Zoom (Board 선택) – 이상 유무 확인 – Close

- BOT Pattern의 Gerber를 확인 후에 저장한다.

Layout 실습 (CAM ③)

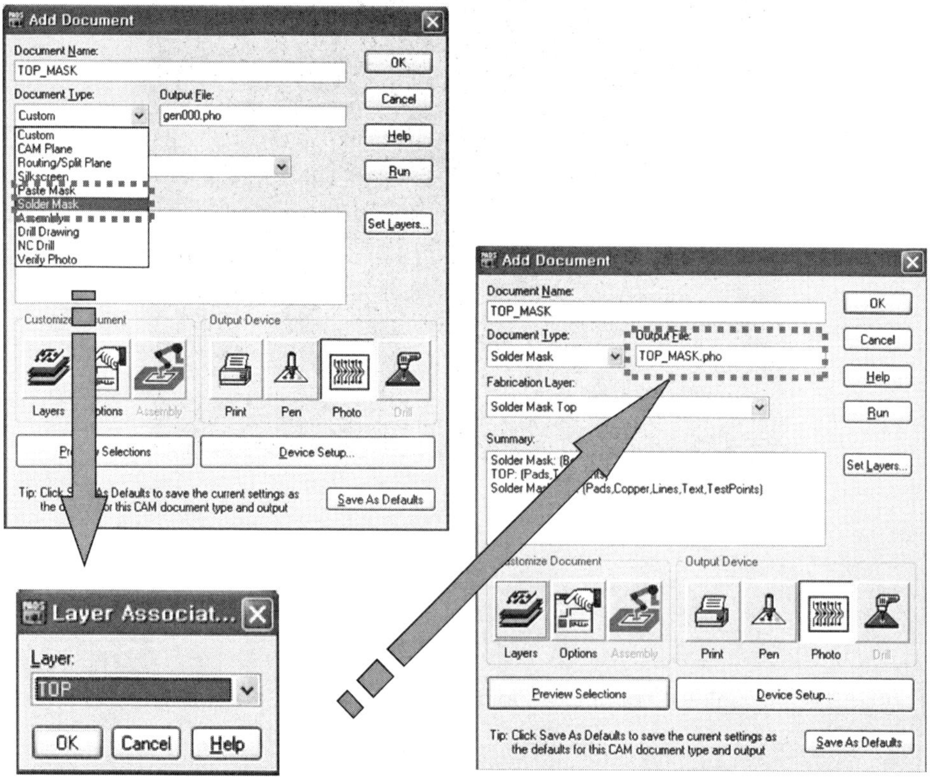

CAM (TOP_MASK 생성)

가. Define CAM Documents – Add – Add Document – Document Name (TOP_MASK) – Document Type (Solder Mask 선택) – Layer Associate – Layer (TOP 선택) – OK – Add Document – Output File (Document Name 동일)

- PCB 제작에 필요한 Gerber DATA는 Output File Name(.pho)이다.

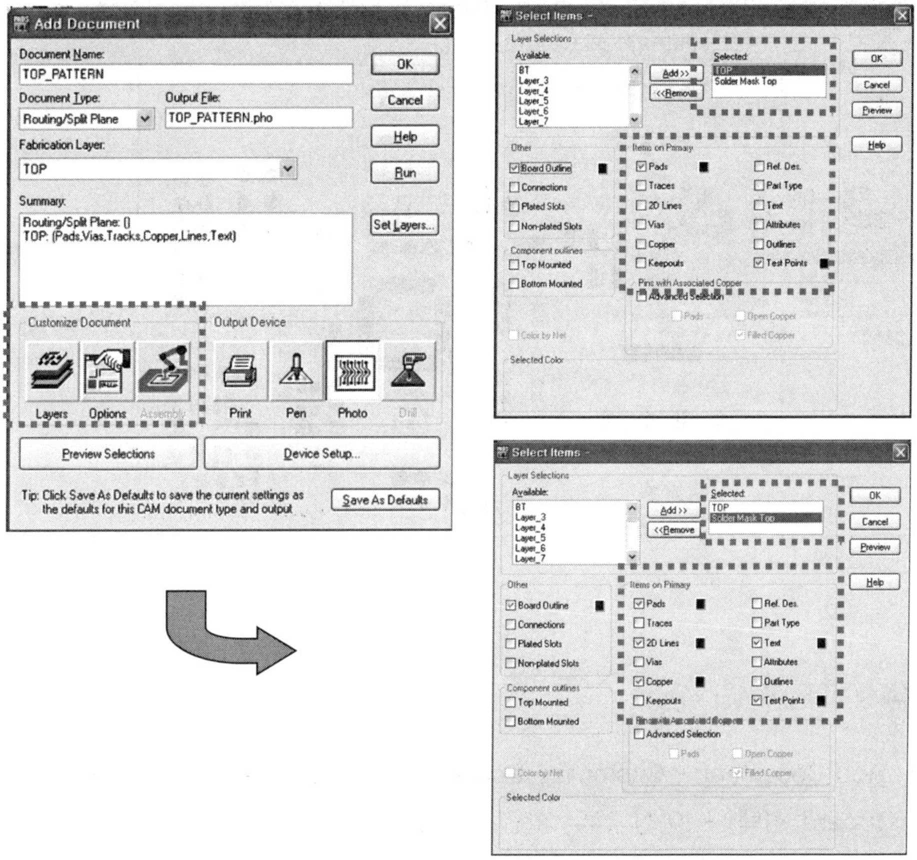

CAM (TOP_MASK 생성)

가. Add Document – Customize Document (Layer 선택) – Other (Board Outline) – Selected (TOP 선택) – Item on Primary (TOP–Pads, Test Points/Solder Mask TOP–Pads, 2D Line, Copper, Text, Text Point 선택 확인) – OK

- 각 Gerber에 해당하는 Item이 선택되어 있는지 확인한다.

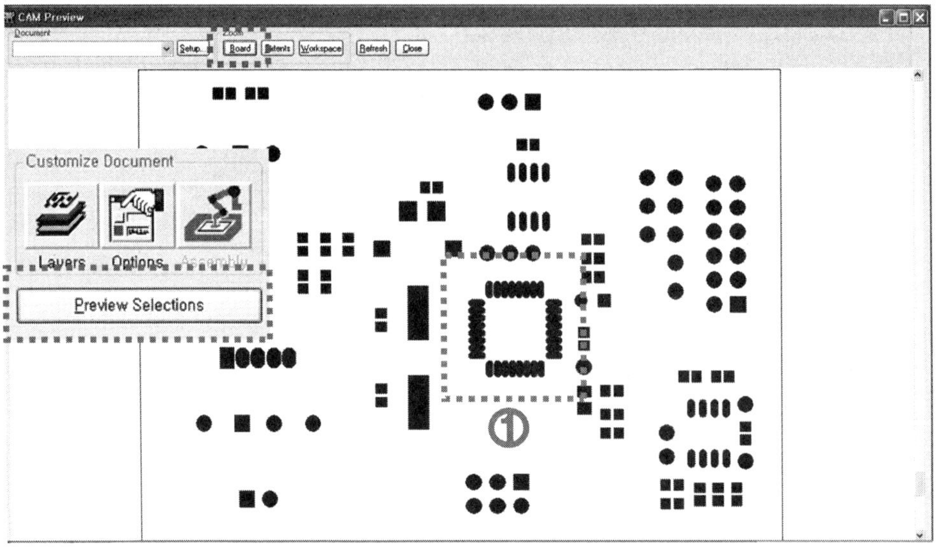

CAM (TOP_MASK 생성)

가. Add Document – Customize Document (Preview Selections 선택) – Zoom (Board 선택) – 이상 유무 확인 – Close

- TOP_Mask의 Gerber를 확인 후에 저장한다.
- ①의 그림처럼 Pads를 1대1로 표현되지 않으면 Option을 변경해야 한다.

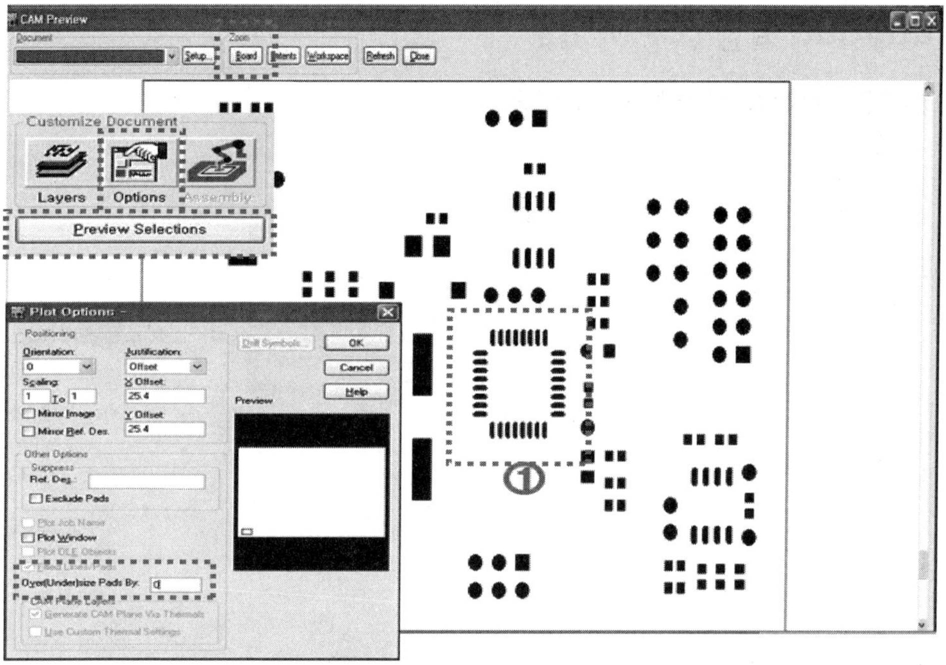

CAM (TOP_MASK 생성)

가. Add Document – Customize Document (Option 선택) – Plot Options – Over size Pads by (0) – OK

- ①의 그림처럼 Pads를 1대1로 표현하려면 Option을 0으로 변경해야 한다.

Layout 실습 (CAM ④)

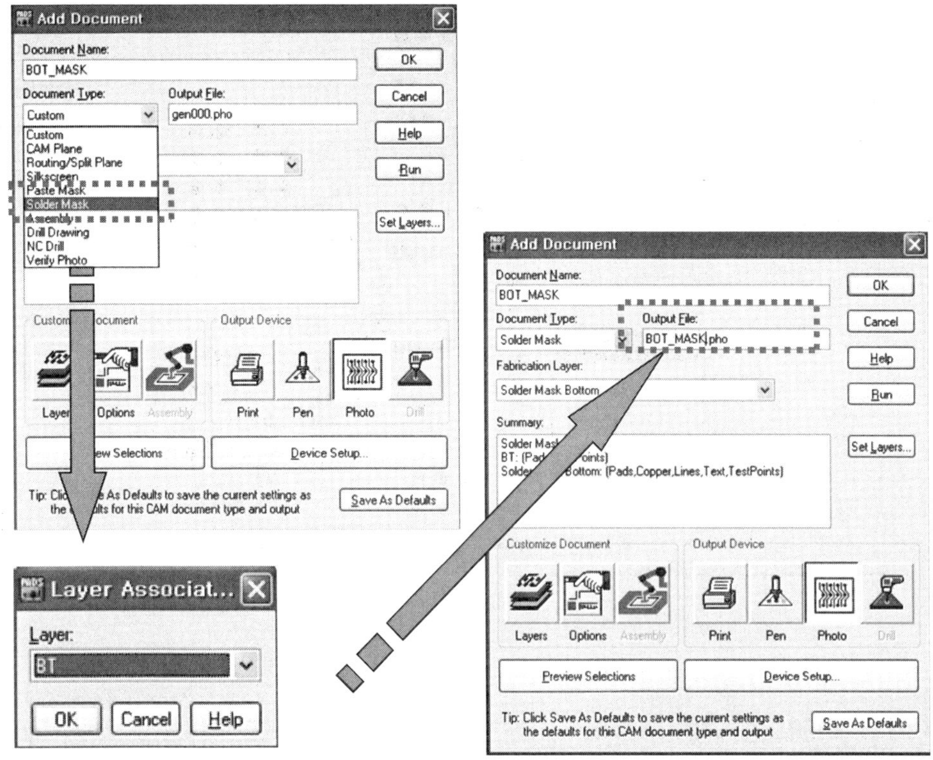

CAM (BOT_MASK 생성)

가. Define CAM Documents – Add – Add Document – Document Name (BOT_MASK) – Document Type (Solder Mask 선택) – Layer Associate – Layer (BOT 선택) – OK – Add Document – Output File (Document Name 동일)

- PCB 제작에 필요한 Gerber DATA는 Output File Name(.pho)이다.

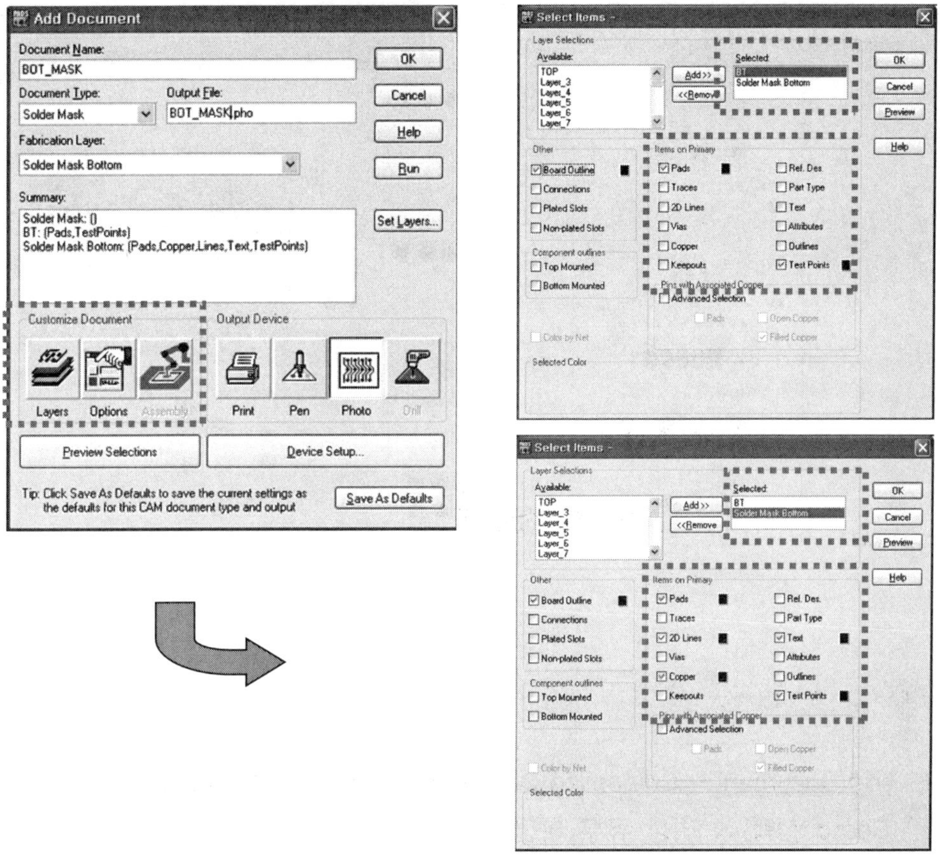

CAM (BOT_MASK 생성)

가. Add Document – Customize Document (Layer 선택) – Other (Board Outline) – Selected (BOT 선택) – Item on Primary (BOT–Pads, Test Points/Solder Mask BOT–Pads, 2D Line, Copper, Text, Text Point 선택 확인) – OK

- 각 Gerber에 해당하는 Item이 선택되어 있는지 확인한다.

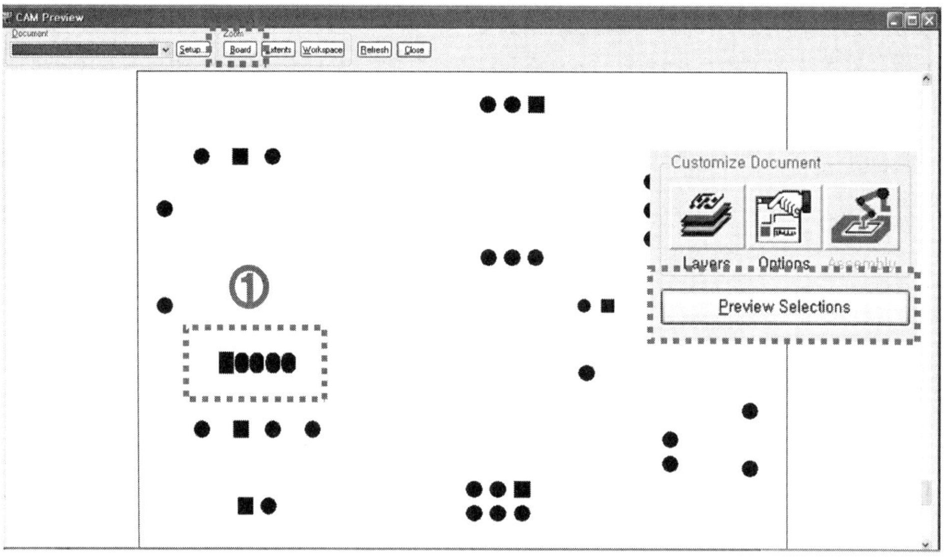

CAM (BOT_MASK 생성)

가. Add Document – Customize Document (Preview Selections 선택) – Zoom (Board 선택) – 이상 유무 확인 – Close

- TOP_ Mask의 Gerber를 확인 후에 저장한다.
- ①의 그림처럼 Pads를 1대1로 표현되지 않으면 Option을 변경해야 한다.

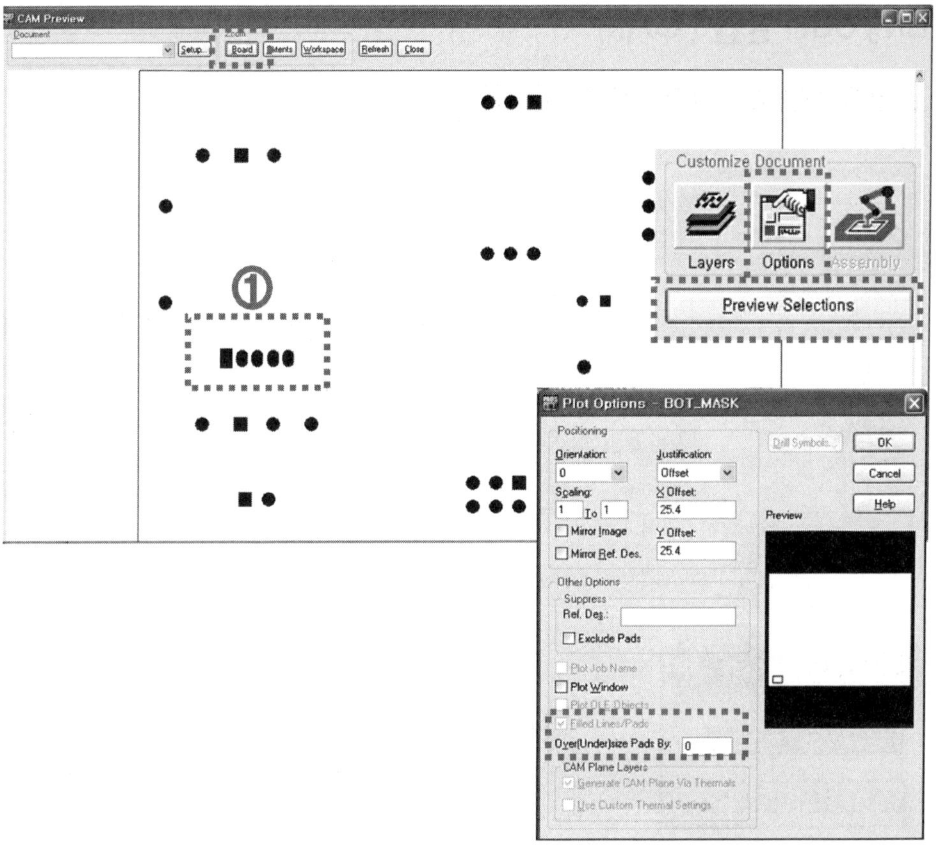

CAM (BOT_MASK 생성)

가. Add Document – Customize Document (Option 선택) – Plot Options – Over size Pads by (0) – OK

- ①의 그림처럼 Pads를 1대1로 표현하려면 Option을 0으로 변경해야 한다.

Layout 실습 (CAM ⑤)

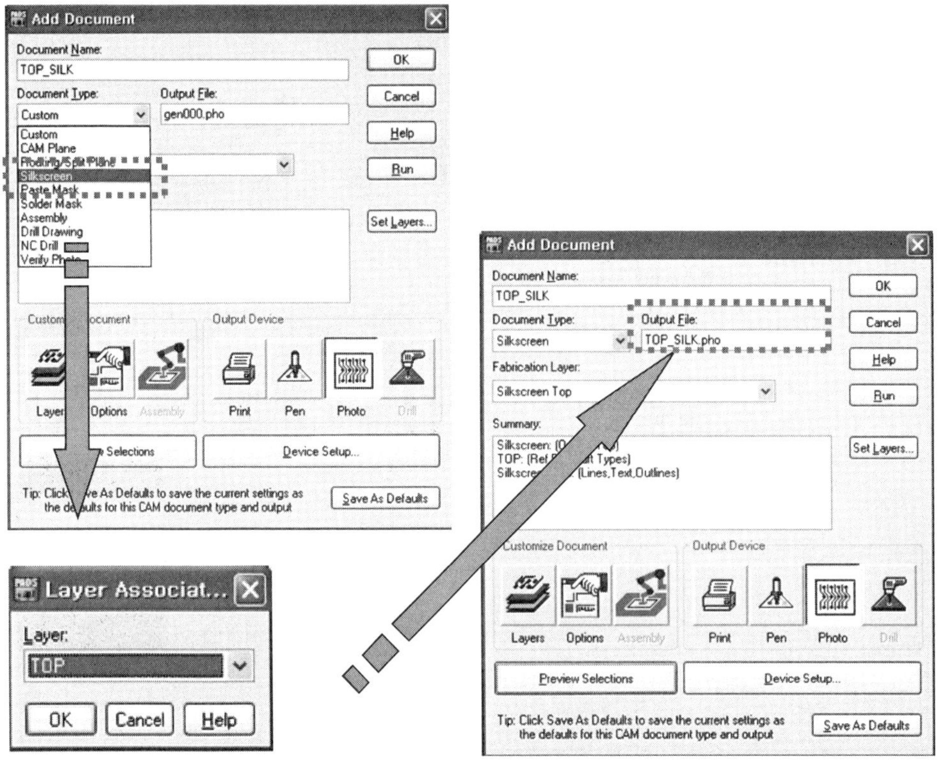

CAM (TOP_Silkscreen 생성)

가. Define CAM Documents – Add – <u>Add Document</u> – Document Name (TOP_SILK) – Document Type (Silkscreen 선택) – Layer Associate – Layer (TOP 선택) – OK – <u>Add Document</u> – Output File (Document Name 동일)

- PCB 제작에 필요한 Gerber DATA는 Output File Name(.pho)이다.

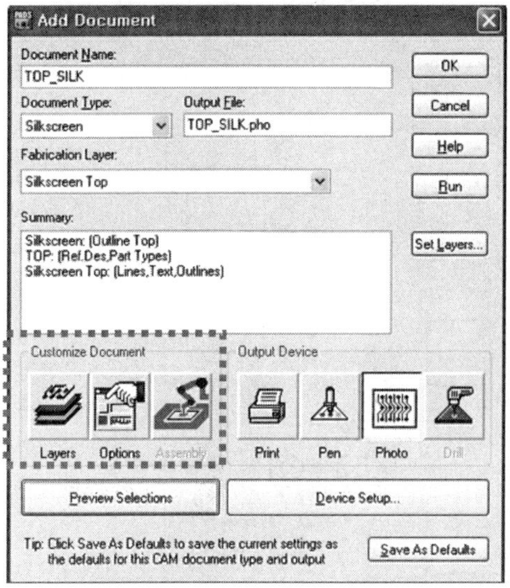

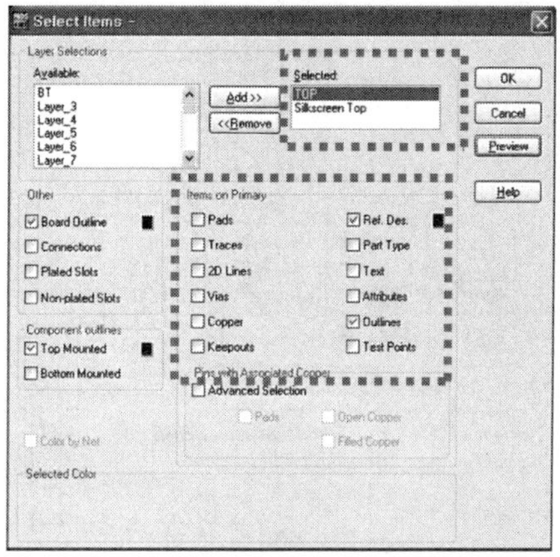

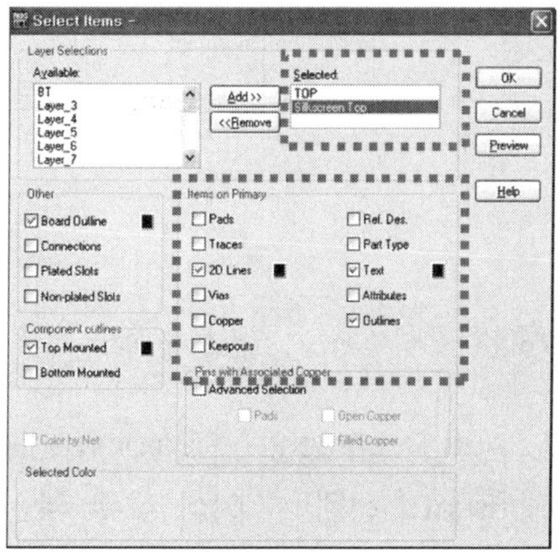

CAM (TOP_Silkscreen 생성)

가. Add Document – Customize Document (Layer 선택) – Other (Board Outline) – Selected (TOP 선택) – Item on Primary (TOP–Ref. Des, Outline/Silkscreen TOP–2D Line, Text, Outline 선택 확인) – OK

- 각 Gerber에 해당하는 Item이 선택되어 있는지 확인한다.

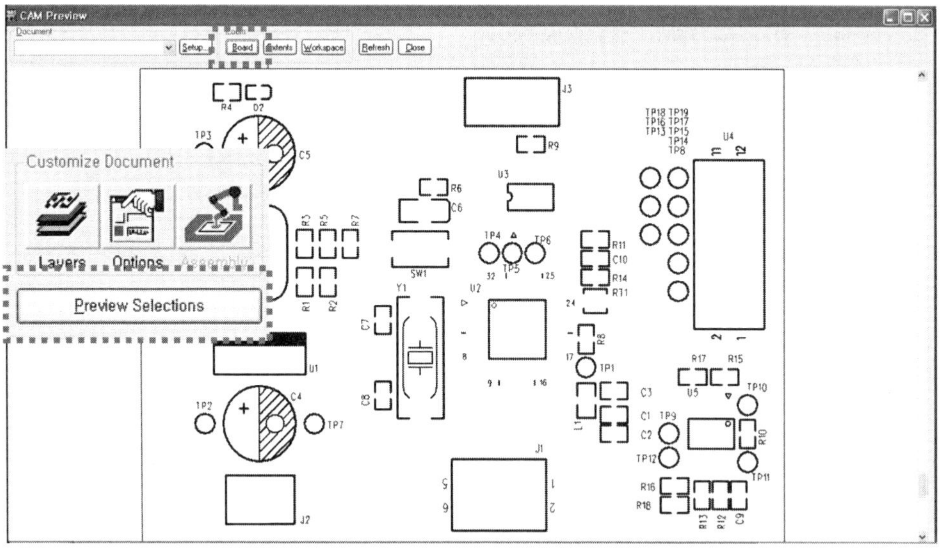

CAM (TOP_Silkscreen 생성)

가. Add Document – Customize Document (Preview Selections 선택) – Zoom (Board 선택) – 이상 유무 확인 – Close

- TOP Silkscreen의 Gerber를 확인 후에 저장한다.

Layout 실습 (CAM ⑥)

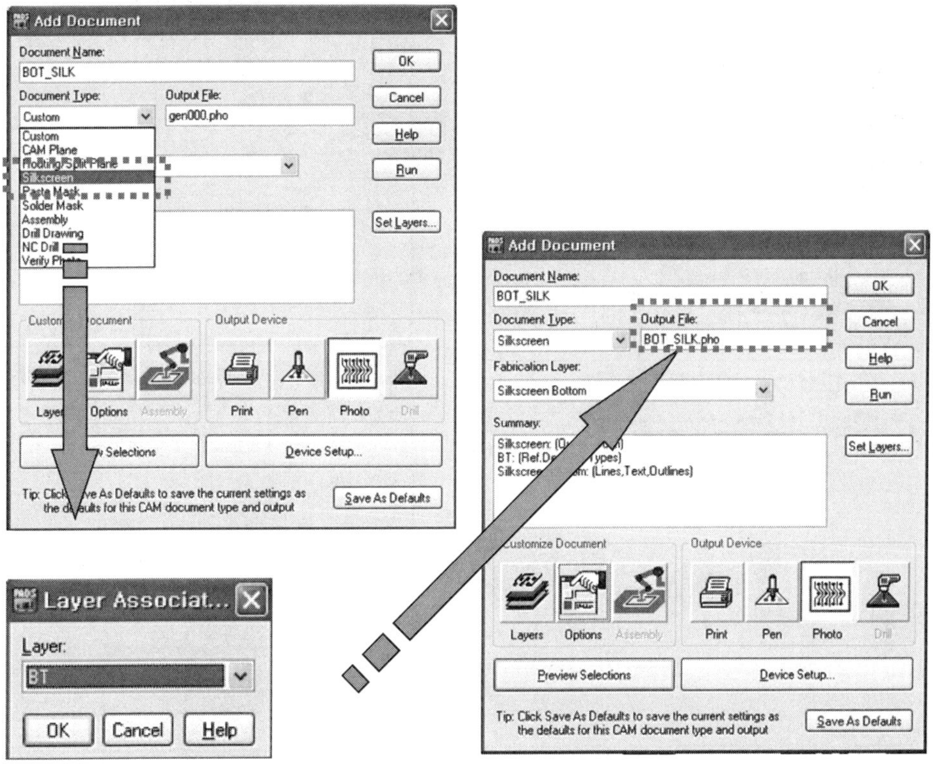

CAM (BOT_Silkscreen 생성)

가. Define CAM Documents – Add – <u>Add Document</u> – Document Name (BOT_SILK) – Document Type (Silkscreen 선택) – Layer Associate – Layer (BOT 선택) – OK – <u>Add Document</u> – Output File (Document Name 동일)

- PCB 제작에 필요한 Gerber DATA는 Output File Name(.pho)이다.

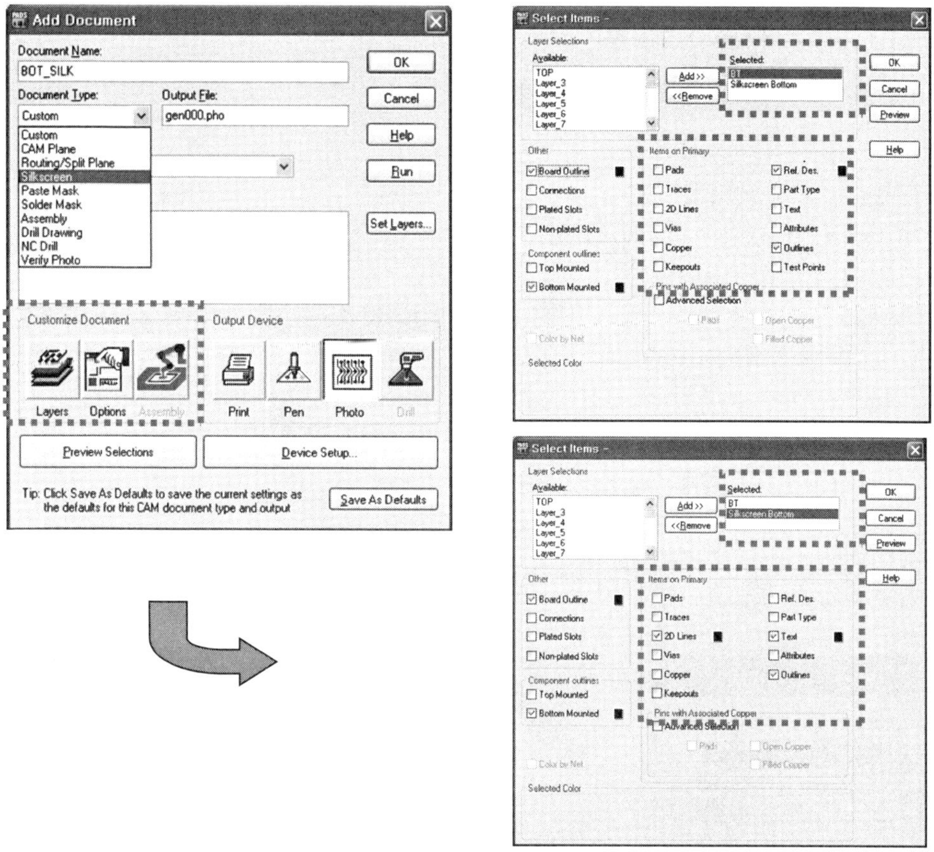

CAM (BOT_Silkscreen 생성)

가. Add Document – Customize Document (Layer 선택) – Other (Board Outline) – Selected (BOT 선택) – Item on Primary (BOT–Ref. Des, Outline/Silkscreen BOT–2D Line, Text, Outline 선택 확인) – OK

- 각 Gerber에 해당하는 Item이 선택되어 있는지 확인한다.

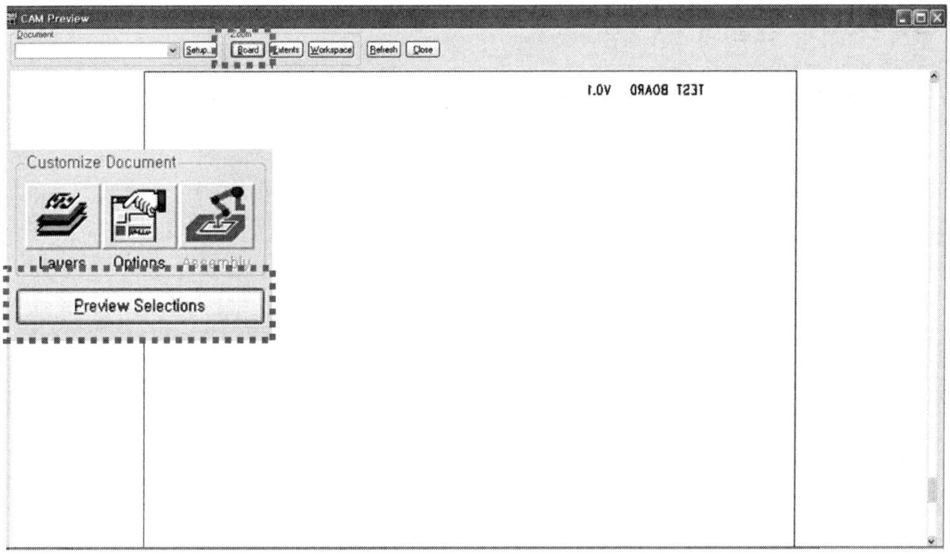

CAM (BOT_Silkscreen 생성)

가. Add Document – Customize Document (Preview Selections 선택) – Zoom (Board 선택) – 이상 유무 확인 – Close

- BOT Silkscreen의 Gerber를 확인 후에 저장한다.

Layout 실습(CAM ⑦)

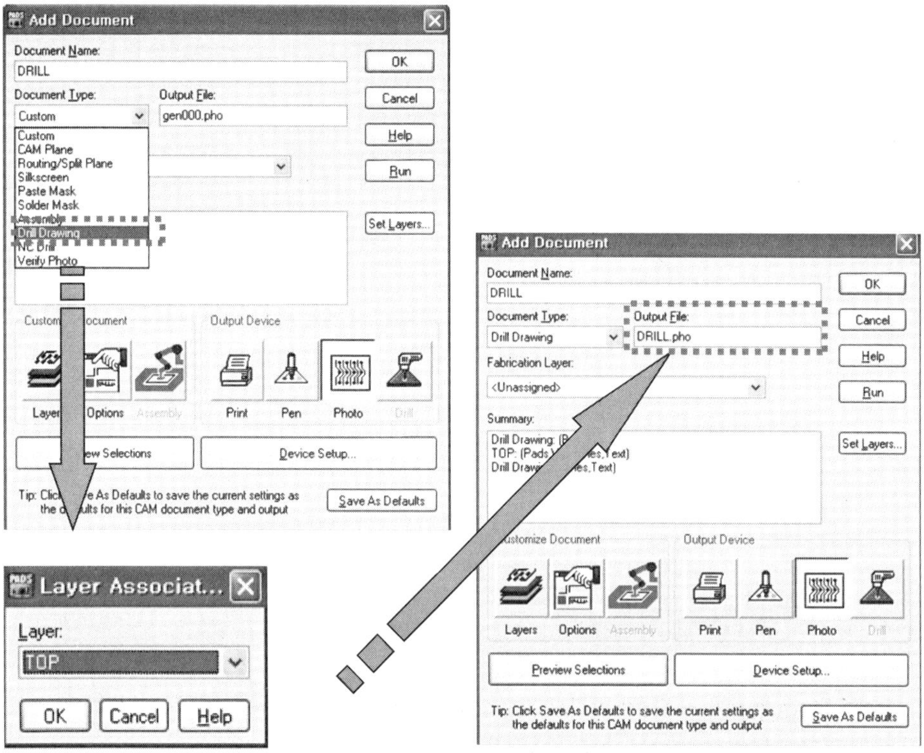

CAM (Drill_Drawing 생성)

가. Define CAM Documents – Add – Add Document – Document Name(Drill) – Document Type(Drill Drawing 선택) – Layer Associate – Layer(TOP 선택) – OK – Add Document – Output File(Document Name 동일)

- PCB 제작에 필요한 Gerber DATA는 Output File Name(.pho)이다.

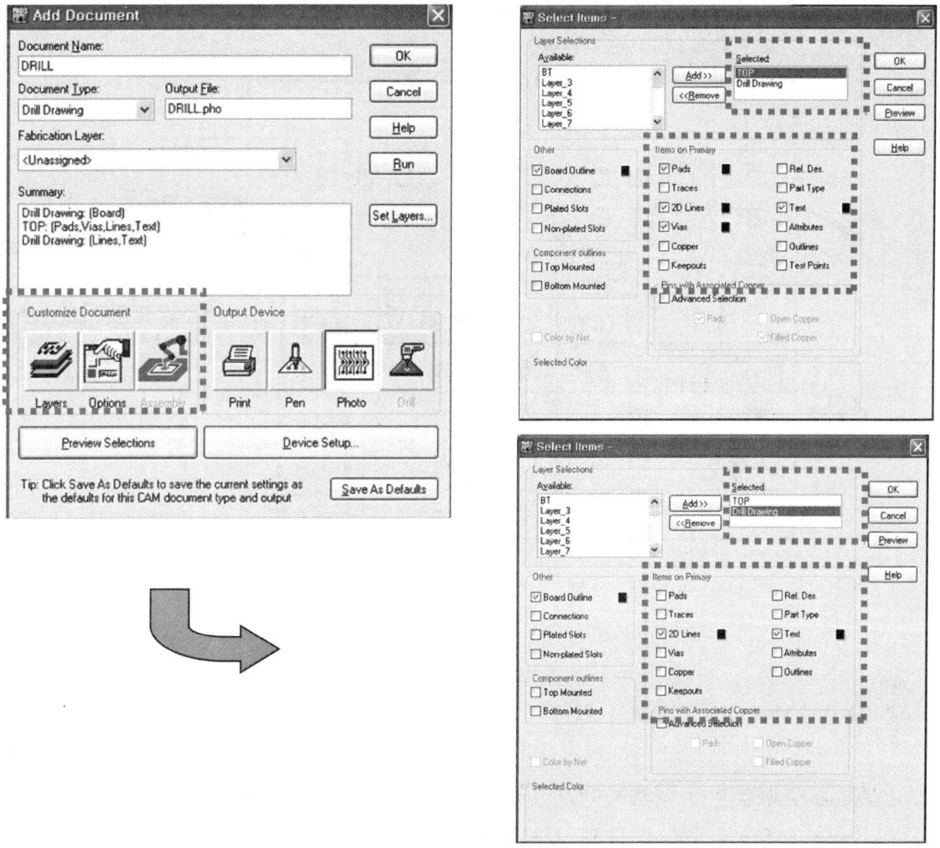

CAM (Drill_Drawing 생성)

가. Add Document – Customize Document (Layer 선택) – Other (Board Outline) – Selected (TOP 선택) – Item on Primary (TOP–Pads, 2Dline, Vias, Text/Drill Drawing–2D Line, Text) – OK

- 각 Gerber에 해당하는 Item이 선택되어 있는지 확인한다.

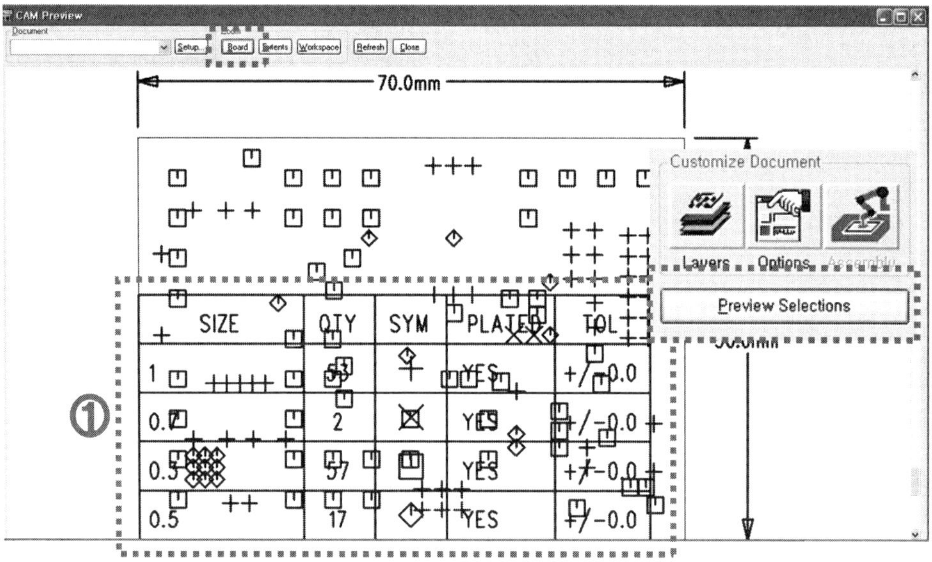

CAM (Drill_Drawing 생성)

가. Add Document – Customize Document (Preview Selections 선택) – Zoom (Board 선택) – 이상 유무 확인 – Close

- TOP_Mask의 Gerber를 확인 후에 저장한다.
- ①의 그림처럼 PCB와 Drill Chart가 겹치면 Option에서 변경하면 된다.

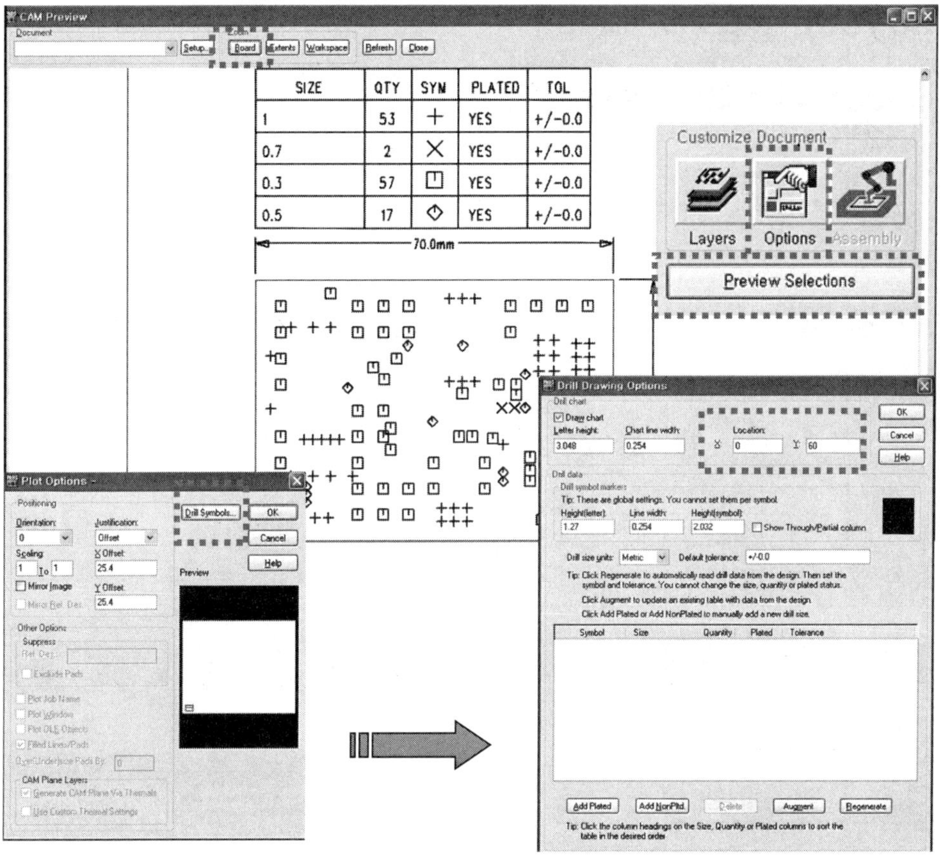

CAM (Drill_Drawing 생성)

가. Add Document – Customize Document (Option 선택) – Plot Options – Drill Symbol 선택 – Location(X/Y=0/60) – OK

- PCB를 기준으로 Drill Chart의 위치를 수정할 수 있다.

Layout 실습 (CAM ⑧)

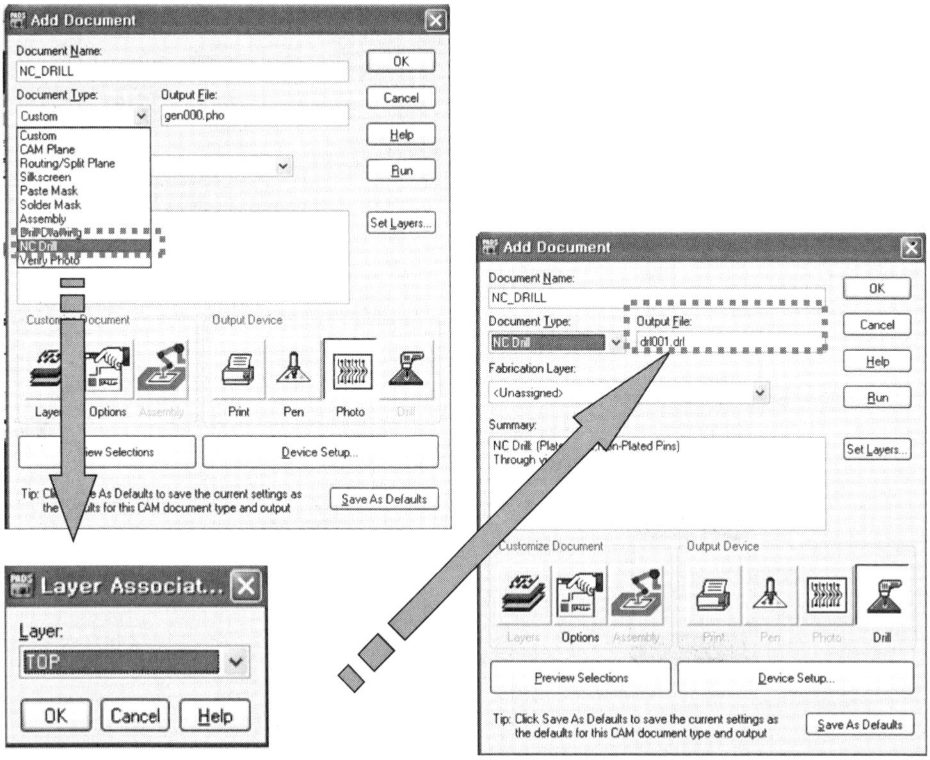

CAM (NC_Drill 생성)

가. Define CAM Documents – Add – <u>Add Document</u> – Document Name (NC Drill) – Document Type (NC_Drill 선택) – Layer Associate – Layer (TOP 선택) – OK – <u>Add Document</u> – Output File (Document Name 동일)

- PCB 제작에 필요한 Gerber DATA는 Output File Name(.pho)이다.

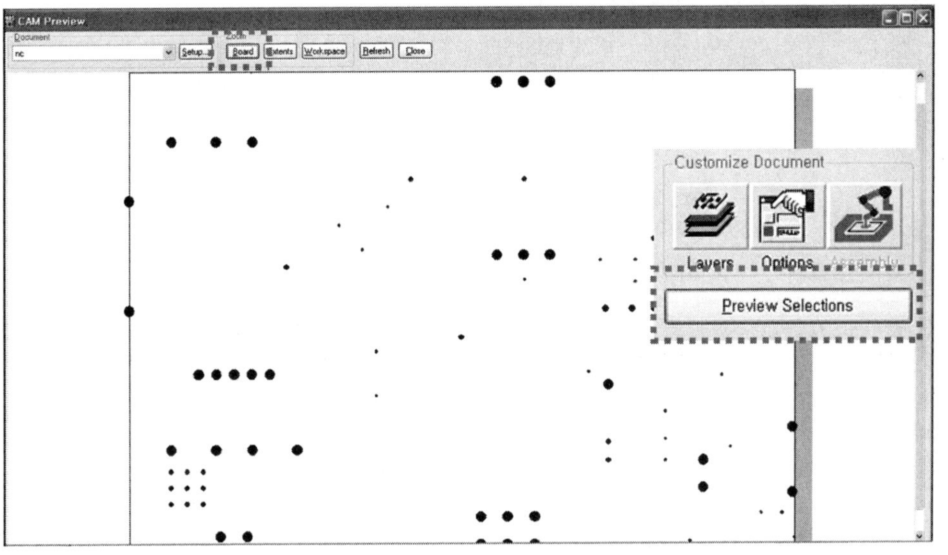

CAM (NC_Drill 생성)

가. Add Document – Customize Document (Preview Selections 선택) – Zoom (Board 선택) – 이상 유무 확인 – Close

- NC Drill의 Gerber를 확인 후에 저장한다.

09 / Layout 실습 (CAM 이해) (2)

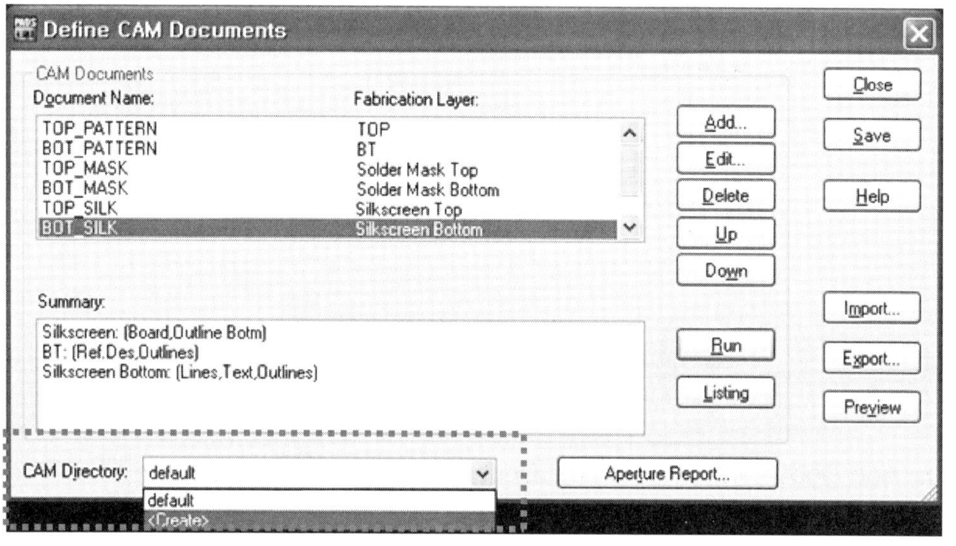

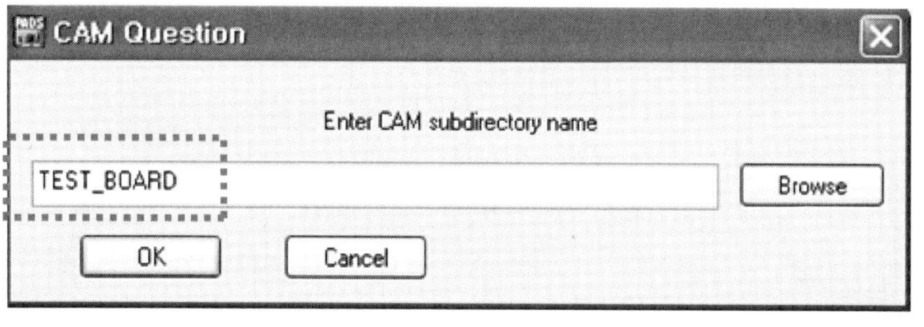

CAM (Gerber 경로지정)

가. Define CAM Documents – Cam Directory (Create) – CAM Question (PCB Name) – OK

- CAM을 이용하여 지정된 경로의 Folder에 Gerber가 저장된다.

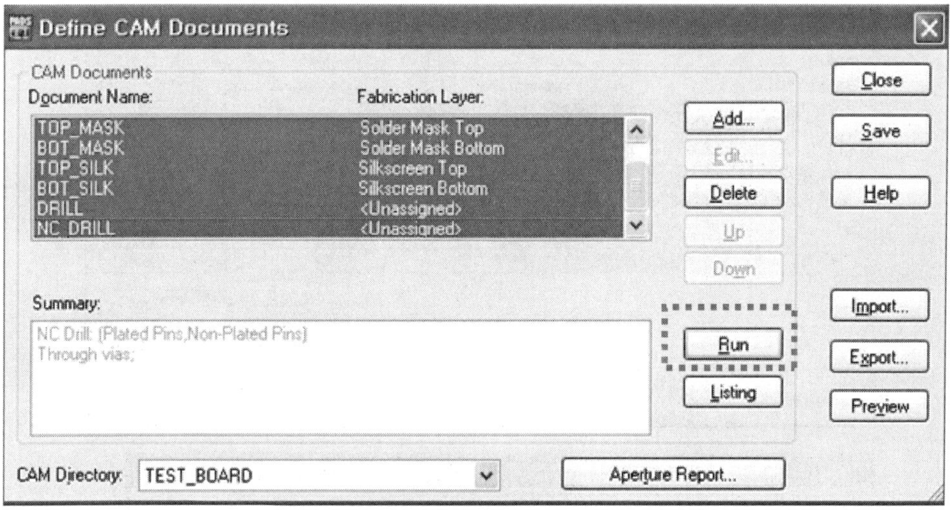

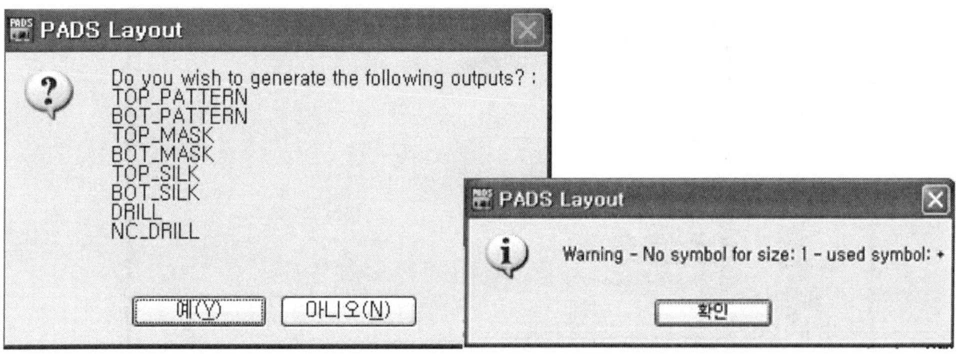

CAM (Gerber 생성)

가. Define CAM Documents -Cam Document (All선택) -Run – PADS Layout (예) – Drill Symbol 확인 -Save – Close

- 전체 Gerber를 생성하는 것과 각각의 Gerber를 생성하는 것이 가능하다.

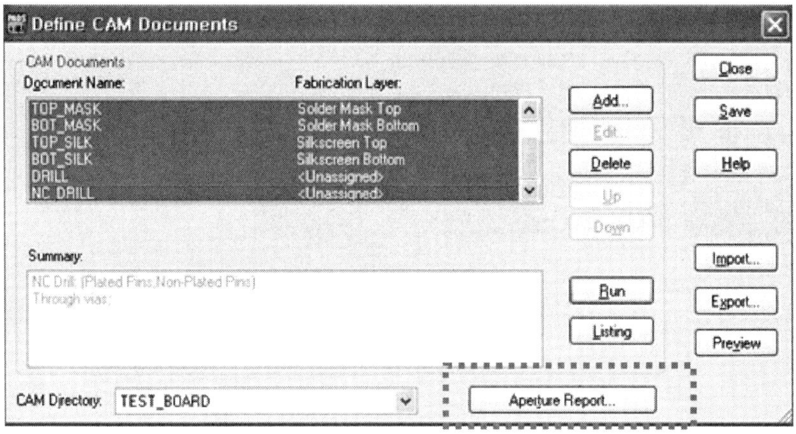

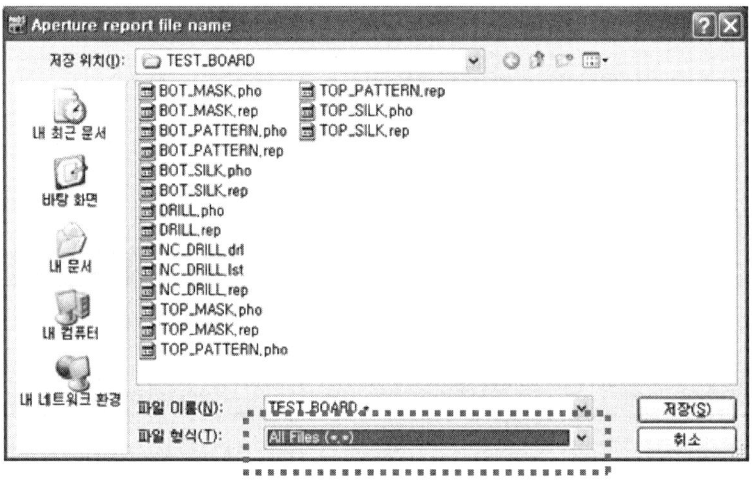

CAM (Gerber 생성)

가. Define CAM Documents -Aperture report file name -파일형식(All file선택)

- Gerber 파일 확인

• 양면 PCB, 양면배치일 경우 기본 Gerber 파일은 총 17개이다.

10 / HOT KEY

PADS 단축키

컨트롤 단축키

[Ctrl+ A]	전체선택
[Ctrl+ U]	하이라이트 OFF
[Ctrl+ B]	보드 아웃라인 표시
[Ctrl+ E]	extent와 동일 기능
[Home key]	보드 전체보기
[Ctrl+ V]	붙여 넣기
[Ctrl+ C]	복사
[Ctrl+ W]	줌 모드 ON/OFF
[Ctrl+ D]	Redraw
[Ctrl+ X]	삭제
[Ctrl+ E]	이동
[Ctrl+ F]	레이어 변경 (Top〈–〉Bottom)
[Ctrl+ Z]	되돌리기
[Ctrl+ G]	그룹 생성 (Create union)
[Alt, S, C] [Ctrl+ Alt+ C]	Display colors setup 대화 상자 오픈
[Ctrl+ Enter]	Option 대화 상자 오픈
[Ctrl+ H]	하이라이트 ON
[Ctrl+ I]	부품 회전 (각도 설정)
[Ctrl+ J]	Route loop
[Ctrl+ Alt+ F]	Selection filter 대화 상자 오픈
[Ctrl+ L]	부품 정렬
[Ctrl+ Alt+ J]	Add Jumper mode
[Ctrl+ M]	Length minimization
[Ctrl+ N]	New file
[Ctrl+ Alt+ N]	View Net 대화 상자 오픈

컨트롤 단축키

[Ctrl+ O]　　　　파일열기
[Alt+ P]　　　　이전 화면 표시 (Toggle)
[Ctrl+ Alt+ S]　　Status Window 오픈
[Ctrl+ Q]　　　　속성
[Ctrl+ R]　　　　반 시계 방향 회전 (90도)
[Ctrl+ S]　　　　저장
[Ctrl+ Page down] 현재 위치를 기준으로 좌표 계산 표시 (Q와 동일)

PADS 명령어

명령어 [명령을 입력하고 Enter key를 누른다.]

AA : Route 또는 2D line작성 시 임의 각도 지원.
AD : Route 또는 2D line작성 시 대각선 각도 지원.
AO : Route 또는 2D line작성 시 90도 지원.
OH : 외곽선을 고해상도로 표시.
OL : 외곽선을 사각형으로 표시.
C : Plane layer상의 anti-pads와 Thermals를 반전.
PO : Pour outline의 On/Off toggle
D : 현재의 작업 Layer를 On/Off
Q : 커서의 현재위치에서 거리측정 mode로 변환.
DO : outline을 On/Off.
F[s] : 파일의 내용을 화면으로 표시.
R[n]: Real Width의 크기를 지정.
DRI : 설계 중 Route의 Intersection만을 체크.
DRP : 설계 중 Design rule을 항상 체크. (DRC ON)
DRO : 설계 중 Design rule을 체크하지 않음. (DRC OFF)
DRW : 부품 배치 중일 때 Error메시지 출력, Routing 중일 때 Design rule을 체크.
S[n][n] : S100 100 : X100, Y100 좌표로 이동.
Space bar : 마우스 왼쪽 버튼과 동일.
S R1.2 : 저항 R1의 2번 Pin을 이동.
SH : Shove mode의 on/Off Toggle
SR[n]n[n] : X,Y에 대한 마우스 커서로부터 상대위치를 검색.
SRY[n] : Y에 대한 상대위치를 검색.
SS[s] : ssu1 : Move mode에서 U1을 검색. (Ref Des로 검색)
SX[n] : X축의 절대위치로 이동.
SY[n] : Y축의 절대위치로 이동.
G[xx] : G20: Design Grid를 20으로 설정.
GD[xx] : GD20: Display Grid를 20으로 설정.
GR[xx] : Design Grid를 설정.
GP : Grid를 Radial grid로 변경 및 해제.
E : End via와 End no via사이를 Toggle.
SPD : Split/mixed planes에 대한 Plane data를 표시.
SPO : Split/mixed plane에 대한 plane polygon outline을 표시.
SPI : Plane thermal indicator를 표시.

명령어 [명령을 입력하고 Enter key를 누른다.]

HC : 2D-line mode에서 원 그리기.
HH : 2D-line mode에서 직선 그리기.
HP : 2D-line mode에서 다각형 그리기.
HR : 2D-line mode에서 직사각형 그리기.
T : Transparent mode의 ON/Off toggle
VA : 자동으로 Via 선택.
V[NAME] : Via Type 선택. (예:V standard via)
VA : 자동으로 Via선택.
VT : Through hole via 사용.
VP : Partial Via 사용.
L[n] : Layer 선택.
N[s] : s로 지정된 net를 Highlight
N : Highlight 해제.
O[r] : Pad또는 trace의 외각 선으로 표시. (Toggle)
W[n] : Width를 변경. (예:W5는 폭으로 5로 변경)
GV[xx] : Via grid setting
UN[n] : 복수개의 Undo실행. (un5:Undo를 5번 실행)
I : Design database의 integrity 체크. (오류는 삭제됨)

PCB 용어

PCB 용어

1. 인쇄회로기판 (PRINTED CIRCUIT BOARD)
   : 전도 선이나, 신호 선을 사용하여 전기적으로 전자부품을 연결한 기판 전기적으로 전자부품을 연결한 기판
2. 프린트 배선 (PRINTED WIRING)
   : 회로설계의 기본으로써 각 부품을 연결시키는 도체 PATTERN을 절연기판의 표면이나 내부에 형성시키는 배선 혹은 기술
3. 인쇄회로기판 (PRINTED CIRCUIT BOARD)
   : 인쇄회로기판을 형성시킨 판
4. 인쇄 배선 판 (PRINTED BOARD)
   : 인쇄배선(프린트 배선)을 형성시킨 판
5. 프린트기판 (PRINTED BOARD)
   : 인쇄회로기판의 약칭
6. 단면 인쇄회로기판(SINGLE-SIDE PRINTED CIRCUIT BOARD)
   : 한쪽 면만 도체회로가 구성된 인쇄회로기판
7. 양면 인쇄회로기판 (DOUBLE(BOTH) – SIDE PRINTED CIRCUIT BOARD)
   : 기판의 양쪽 면에 도체회로가 구성된 인쇄회로기판.
8. 다층 인쇄회로기판(MULTI LAYER PRINTED CIRCUIT BOARD)
   : 기판의 표면양쪽은 물론이고 내층에도 도체회로로 구성된 인쇄회로기판
9. 플랙시블 배선 판(FLEXIBLE PRINTED CIRCUIT BOARD)
   : 유연성이 좋은 절연성을 이용한 인쇄회로 판
10. 플랙스-리지드 배선 판(FLEX-RIGID PRINTED CIRCUIT BOARD)
   : 유연성 및 경질성의 절연성이 2종류를 명명의 용도 특성을 살린 복합체의 인쇄회로기판.
11. 메탈 코아 배선 판(METAL CORE PRINTED CIRCUIT BOARD)
   : 기재가 금속판으로 된 인쇄회로 판.

PCB 용어

12. 마더 보드(MOTHER-BOARD)

: 인쇄회로의 조립 품을 취급함에 있어서 동시에 접속연결 가능한 회로 판.

13. 부품 면(COMPONENT SIDE)

: 인쇄회로기판에 있어서 대부분의 부품이 탑재되는 면

14. 납땜 면(SOLDER SIDE)

: 부품 면의 반대 면으로 각 부품의 납땜이 이루어지는 면.

15. 격자(格子 : GRID)

: 인쇄회로 판의 설계상 부품의 위치결정시 수직 수평선을 맞춰서 할 수 있도록 제조된 망목

16. 패턴(PATTERN)

: 인쇄회로기판에 형성되는 도전성 또는 비 도전성의 도형

17. 도체패턴(CONDUCTIVE PATTERN)

: 인쇄회로에서 전류가 흐를 수 있는 도전성 재료가 형성된 부분

18. 비 도체 패턴(NON-CONDUCTIVE PATTERN)

: 인쇄회로기판의 회로에서 전류가 통할 수 없는 부분으로 도체PATTERN을 제외한 부분.

19. 도체(CONDUCTOR)

: 도체회로에서 각각의 도전성 금속

20. 플러시 도체(FLUSH CONTACT or CONTACT PATTERN)

: 도체의 표면이 절연성의 표면과 격차간격이 없이 동일한 평면상태의 도체판

21. 컨택 회로(PRINTED CONTACT or CONTACT PATTERN)B

: 전기적 접촉을 목적으로 형성된 회로부분.

22. 에지 커넥터 단자(EDGE BOARD CONDUCTS)

: 인쇄회로기판의 끝부분에 회로 컨택으로 소켓에 연결해 사용토록 설계된 단자.

23. 프린트 부품(PRINTED COMPONENT)

: 인쇄기술에 의해서 형성된 전자부품.

PCB 용어

24. 태핑 부품(TAPED COMPONENT)

: 테이프에 부품을 연속으로 배열하여 사용하기 편리하게 만든 부품.

25. 인쇄회로기판 조립 품(PRINTED BOARD ASSEMBLY : PCB, A'ssy)

: PBA는 인쇄회로기판(PCB: Printed Circuit Board)에 표면실장(SMT) 공정을 거쳐 제품이 완성된 상태.

26. 절연 판(BASE MATERIAL)

: 인쇄회로기판의 재료에 있어서 표면에 회로를 형성하기가 용이한 연결 재료 판.

27. B스테이지 수지(B-STAGE RESIN)

: 경화반응의 중간단계에 있는 열 경화 성 수지.

28. 프리프러그(PREPREG)

: B 스테이지 수지처리가 되어 있는 유리 섬유 포 SHEET의 다층 인쇄 회로 판의 재료.

29. 윅킹현상(WICKING)

: 절연 판 섬유질의 결을 따라 액체의 모관 흡수현상.

30. 본딩시트(BONDIBG SHEET)

: 다층 인쇄 회로 판을 제조할 때 각층간의 접착 시 사용하는 것으로 적당한 접착성이 있는 프리프러그 등의 재료 SHEET.

31. 금속베이스 기판(METAL CLAD BASE MATERIAL)

: BASE의 소재가 금속 등으로 제작된 양면 또는 단면 기판용 절연기판.

32. 동박 적층 판(COPPER CLAD LAMINATED BOARD)

: 절연 판의 한쪽 면 또는 양쪽 면에 동박을 입힌 인쇄회로기판용 적층 판.

33. 동박 호일(COPPER FOIL)

: 절연 판의 한쪽 면 또는 양쪽 면에 접착시켜서 도체회로를 형성하는 얇은 동판으로 원통에 감겨진 상태.

34. 적층 판(LAMINATED)

: 2장 혹은 그 이상의 얇은 절연재료에 수지를 합침 시켜서 적층하여 가압 가열에 의하여 만들어지는 판.

PCB 용어

35. 관통접속(THROUGH CONNECTION)

: 인쇄회로기판의 양쪽 면에 있는 도체를 서로 관통하여 전기적 접속

36. 층간 접속(INTERLAYER CONNECTION)

: 인쇄회로기판의 서로 다른 층과 도체 PATTERN을 전기적 접속

37. 와이어 관통접속(WIRE THROUGH CONNECTION)

: WIRE를 관통시켜 연결 접속시킴.

38. 점퍼와이어(JUMPER WIRE)

: JUMPER용 와이어선

39. 도금 스루 홀(PLATED THROUGH HOLE)

: HOLE의 내벽 면에 금속을 도금하여 전기적 접속 관통시킨 홀.

40. 랜드 리스 홀(LANDLESS HOLE)

: LAND가 없는 도금 스루 홀

41. 랜드(LAND)

: 제품의 연결 납땜을 위한 도체 PATTERN의 일부분.

42. 엑세스 홀(ACCESS HOLE)

: 다층 회로 판 내층의 표면을 상출 시킬 홀

43. 크리어런스 홀(CLEARANCE HOLE)

: 다층 회로 판에 있어서 도금 스루 홀에 의해 각 층간 전기적 접속을 하고 있으나 층의 중간에서 전기적 접속을 요구하지 않는 곳에 LAND를 삭제한 영역.

44. 위치선정 홀(LOCATION HOLE)

: 정확한 위치를 선정하기 위해 회로 판이나 PANEL에 HOLE를 뚫음.

45. 마운팅 홀(MOUNTING HOLE)

: 회로 판의 기계적 취급이 용이하게 하기 위한 HOLE.

46. 부품 홀(COMPONENT HOLE)

: 회로 판에 부품의 끝 부분이 삽입되어 도체의 LAND와 납땜 등 전기적 접속이 이루어지기 위한 홀.

PCB 용어

47. 홀 패턴(HOLE PATTERN)

: 인쇄 회로 판의 모든 홀의 배치.

48. 크로스 해칭(CROSS-HATCHING)

: 넓은 도체의 PATTERN 중에서 불필요한 도체부분을 없애 도체의 면적을 감소.

49. 플러라이징 슬롯(POLARIZING SLOT)

: 적합한 커넥터에 삽입하였을 때 정확하게 커넥터 단자와 간격을 맞추기 위해 인쇄회로기판의 끝 부분에 정확한 설계

50. 아트웍 마스터(ART WORK MASTER)

: 제조용 원판 필름을 만들기 위하여 만들어진 원도를 만드는 작업공정

51. 제작 데이터(DATUM REFERENCE)

: HOLE 위치, SIZE, 층수 등의 특기사항의 작업, 내용 등을 수록.

52. 랜드 폭(ANNULAR WIDTH)

: 홀의 주위에 있는 도체의 폭.

53. 앤뉴럴 링(ANNULAR RING)

: 랜드에서 홀의 부분을 제외한 나머지 도체부분.

54. 층(LAYER)

: 다층인쇄회로기판을 구성하고 있는 모든 층으로서 기능적인 도체 층, 절연 층, 구성상의 내층, 표면층, 커버층

55. 카바 층(COVERLAYER)

: 인쇄회로기판의 도체회로를 보호하기 위해 입힌 절연 층.

56. 절연보호코팅(CONFORMAL COATING)

: 완성된 인쇄회로기판 조립품의 표면 상태를 보호하기 위한 절연보호피막

57. 바이아 홀(VIA HOLE)

: 다층 인쇄회로판에서 다른 층과 전기적 접속을 좋게 하기 위해 스루 홀 도금을 한 홀.

58. 도체 층간 두께(LAYER-TO-LAYER-SPACING)

: 다층인쇄회로기판에서 인접한 도체층간의 절연 층 재료의 두께

PCB 용어

59. 블라인드 도체(BLIND CONDUCTOR)

: 인쇄회로기판에 있어서 전기적 특성과 관계없는 도체.

60. 작업지도서(ROAD MAP)

: 인쇄회로기판의 조립 및 수리작업이 편리하도록 회로 및 부품의 형태를 비도전 재료에 인쇄

61. 심볼 마크(SYMBOL MARK)

: 인쇄회로기판의 조리 및 수리작업을 편리하도록 부품의 형태 및 기호를 표시.

62. 제조도면(MANUFACTURING DRAWING)

: 인쇄회로기판의 사양, 특성, 외형, PATTERN의 배치, 홀 SIZE 등의 규정을 표기한 도면.

63. 사진축소(PHOTOGRAPHIC REDUCTION DIMENSION)

: 원도를 축소 촬영하는 공정.

64. 포지티브(POSITIVE)

: 양획, PATTERN 부분의 불투명한 상태.

65. 포지티브 패턴(POSITIVE PATTERN)

: 도체회로부분이 불투명한 상태.

66. 네가티브(NEGATIVE)

: 음획, 포지티브의 반대로서 회로부분이 투명한 상태.

67. 네가티브 패턴(NEGATIVE PATTERN)

: 도체회로부분이 투명한 상태.

68. 제조용 원판(ORIGINAL PRODUCTION MASTER)

: 제조용 원판 FILM을 양산용으로 여러 장 복사 편집한 원판 필름.

69. 판넬

: 양산 작업 시 1매 이상 편집된 크기의 원자재.

70. 양산용 원판(MULTIPLE IMAGE PRODUCTION MASTER)

: 제조용 원판 필름을 양산용으로 여러 장 복사 편집한 원판 필름.

PCB 용어

71. 양산용 패턴(MULTIPLE PATTERN)

: 한 장의 패널의 크기 내에 2매 이상 편집된 패턴.

72. 양산용 인쇄 패널(MULTIPLE PRINTED PANEL)

: 양산용 원판 필름으로부터 인쇄가 종료된 패널.

73. 서브 트랙티브 법(SUBTRACTIVE PROCESS)

: 원자재인 동박 적층 판에서 필요한 도체의 회로부분을 제외한 불필요한 부분의 동 또는 금속을 제거시켜 필요한 도체회로를 형성하는 인쇄회로 기판의 공법.

74. 에디티브법(ADDITIVE PROCESS)

: 절연 판에 도전성 재료를 이용해서 필요한 도체회로를 직접 형성 시키는 공법.

75. 폴 에디티브 법(FULLY ADDITIVE PROCESS)

: 절연 판에 무전해 도금과 전해도금을 이용해서 필요한 도체회로를 형성한 공법.

76. 세미 에디티브 법(SEMI ADDITIVE PROCESS)

: 절연 판에 무전해 도금과 전해도금 법 다시 에칭 등의 방법을 이용하여 필요한 도체회로를 형성시키는 제조공법.

77. 레지스트(RESIST)

: 제조 및 시험 공정 중 에칭 액, 도금액, 납땜 등의 특정한 영역에서 보호용으로 사용되는 피복재료.

78. 에칭(ETCHING)

: 도체회로를 얻기 위하여 원자재에서 불필요한 금속부분을 화학적 혹은 전기적 방법으로 제거하는 공법.

79. 에찬트(ETCHANT)

: 에칭에 사용되는 부식성 용액.

80. 포토에칭(PHOTO ETCHING)

: 원자재인 금속박판상에 감광성 레지스트를 도포한 후 사진제판법에 의해 불필요한 부분을 에칭하는 공정.

PCB 용어

81. 차별에칭(DIFFERENTIAL ETCHING)

: 에칭의 도금 두께를 조절하기 위해 필요한 두께 외의 부분을 부분 에칭하는 공정.

82. 에칭 펙터(ETCHING FACTOR)

: 도체두께 방향의 에칭 깊이를 도체 폭 방향의 깊이의 대비.

83. 네일 헤드(NAIL HEADING)

: 다층 인쇄 회로 판에서 내층회로의 드릴링 홀 부분에 못의 머리 모양으로 도금이 되어 있는 부분.

84. 홀 세정(HOLE CLEANING)

: 용내의 도체표면을 깨끗이 청정하게 세정.

85. 도금(PLATING)

: 화학적 혹은 전기화학적 반응으로 절연 판, 스루 홀, 도체 패턴에 금속을 도금.

86. 패널 도금(PANEL PLATING)

: 패널 전표면(HOLE 포함)에 하는 도금.

87. 패턴도금

: 도체 패턴 부분에 선택적인 도금.

88. 텐팅법(TENTING)

: 도금 스루 홀에서 도체 패턴을 레지스트 (통상은 드라이필름)에 의한 에칭법.

89. 도금 리드(PLATING BAR)

: 전기도금에서 상호 접속시키는 금속.

90. 오버 도금(OVER PLATING)

: 형성된 도체회로에 필요이상의 금속이 도금된 것.

91. 솔더 레지스트(SOLDER RESIST)

: 인쇄회로기판의 특정영역에 도포시키는 재료로써 납땜 작업 시 부분 납땜이 가능하며 납땜이 불필요한 부분의 회로도 보호.

PCB 용어

92. 스크린 인쇄(SCREEN PRINTING)

: 제판이 되어 있는 스크린 원판에 잉크를 붓고 스퀴지로 압력을 가하면서 인쇄하여 판넬의 표면에 모양을 전사시키는 방법.

93. 휴징(FUSING)

: 도체 회로의 SOLDER 금속 피복을 용해시켜 다시 응고해 깨끗하게 하는 공정.

94. 핫 에어 레벨링(HOT AIR LEVELLING)

: 열풍으로 필요 없는 부분의 납을 용해 제거하면서 납 표면을 고르게 하는 방법.

95. 솔더 레벨링(SOLDER REVELLING)

: 회로표면의 납을 기계적인 방법의 열, 열풍 등의 방법으로 표면의 광택 작업.

96. 플로우 솔더(FLOW SOLDER)

: 용해된 납이 탱크에서 상하순환 하면서 용해된 납의 표면이 회로 판에 납땜되는 방법.

97. 딥 솔더(DIP SOLDER)

: 부품이 삽입된 회로 판을 용해된 납조의 정지 표면에 접촉하면서 납땜 방법.

98. 웨이브 솔더링(WAVE SOLDERING)

: FLOW SOLDERING과 동일.

99. 기상 솔더링(VAPOUR PHASE SOLDERING)

: 증기 에네르기가 방출되면서 납을 용해시켜 납땜 공정을 이행하는 방법.

100. 리플로우 솔더링(REFLOW SOLDERING)

: 납땜을 하고자 하는 부품에만 납을 용해 시켜서 납땜하는 작업.

11 / 실습예제

1. CAE Decal 예제

예제 1) CON2P

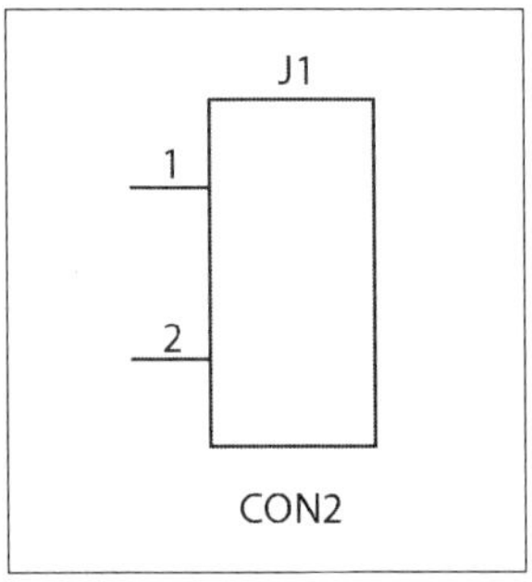

예제 2) CON3P

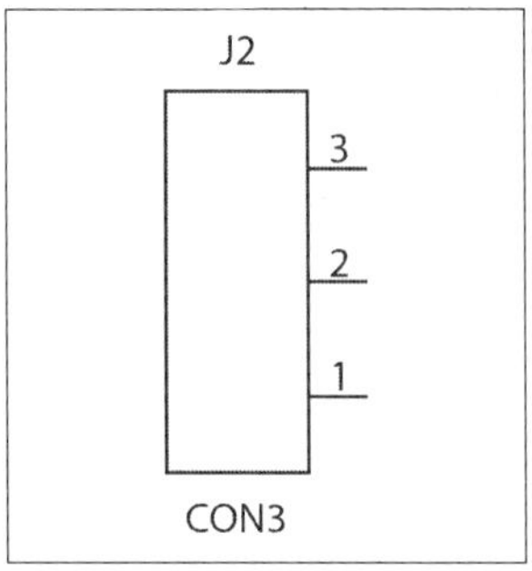

2. CAE Decal 실습

예제 1) 1N914

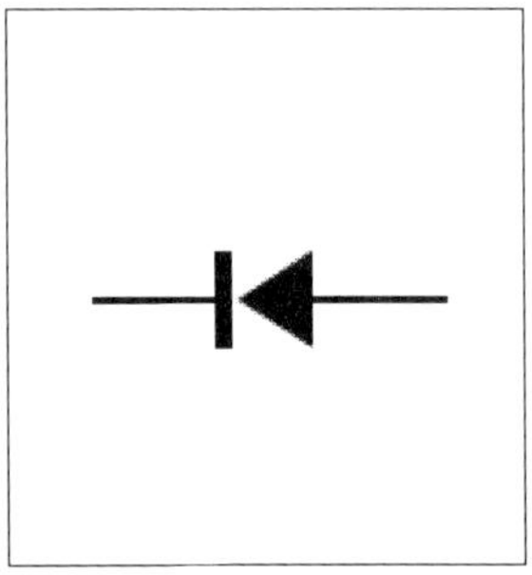

예제 2) LM2575

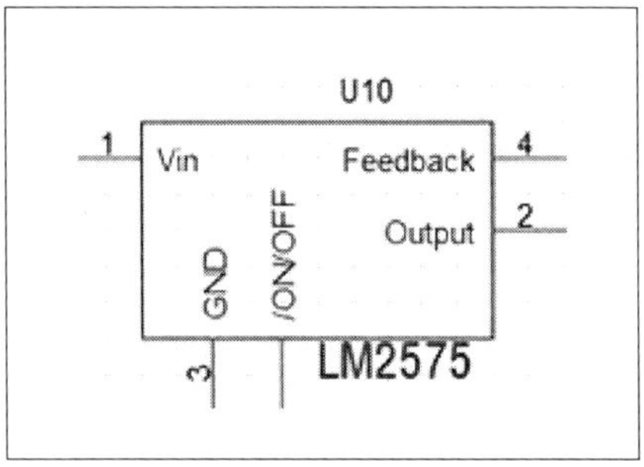

+3. 수동 PCB Decal

예제 1) 전해 cap

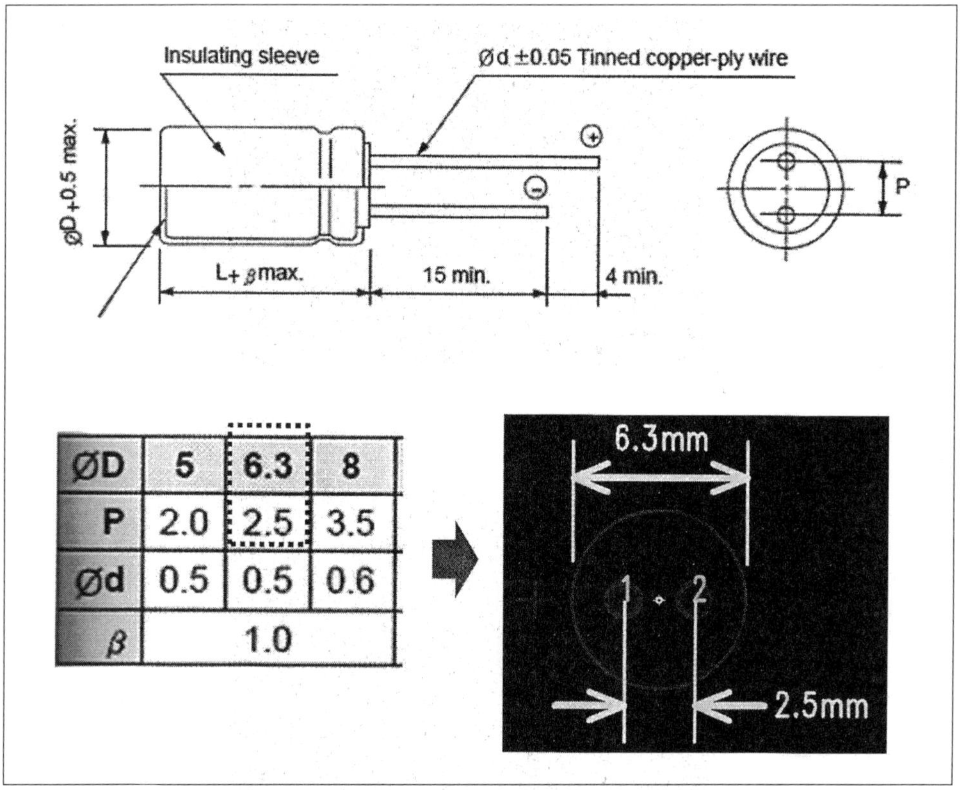

예제 2) 10pin connector

인터넷 부품 검색해 보기

예제 3) 저항 R1608

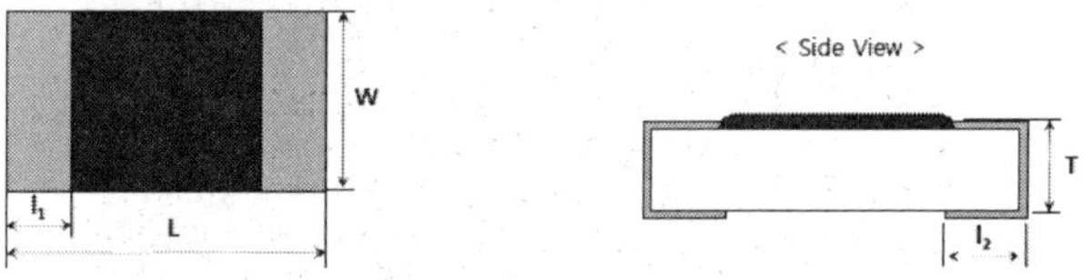

[Unit : mm]

TYPE (Inch)	L	W	T	l_1	l_2	Unit Weight
0402 (01005)	0.40±0.02	0.20±0.02	0.13±0.02	0.10±0.03	0.10±0.03	0.04mg
0603 (0201)	0.60±0.03	0.30±0.03	0.23±0.03	0.10±0.05	0.15±0.05	0.15mg

4. Wizard PCB Decal

예제 1) DUAL-Dip14

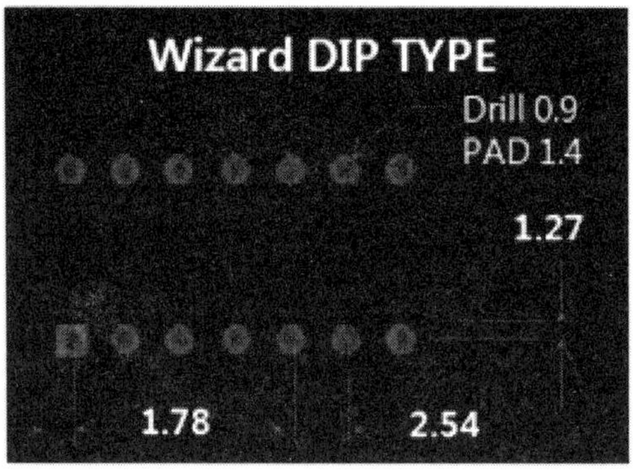

예제 2) DUAL-SMD(SO14)

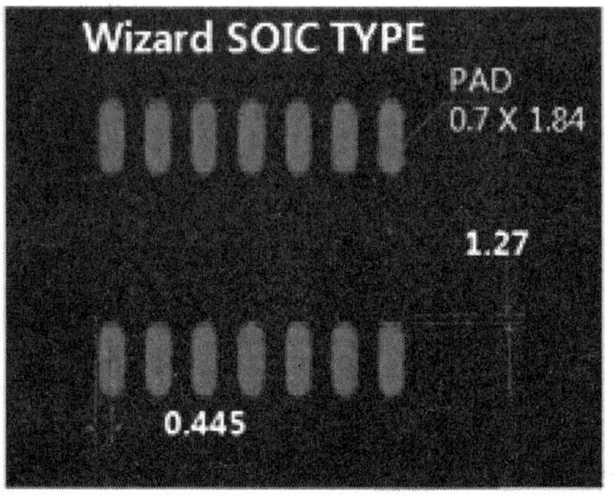

예제 3) QUAD-QFP32

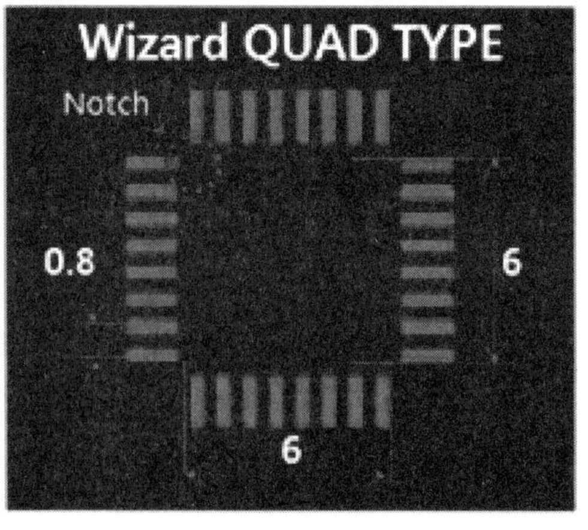

5. Copper 이용한 PCB Decal

예제 1) 다각형 PAD(Associate)

예제 2) KEY-PAD(Combine)

6. PCB Decal (PAD size)

예제 1) SMB

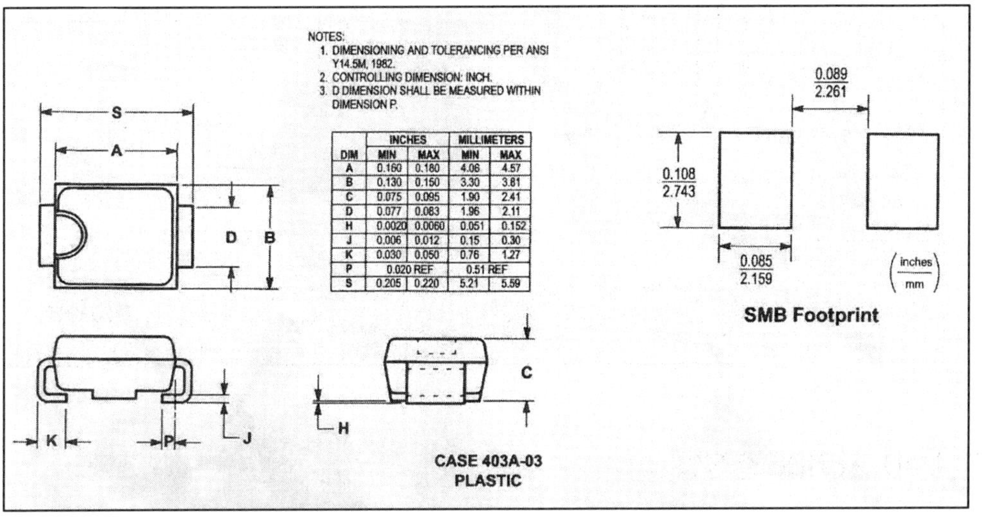

DIM	INCHES MIN	INCHES MAX	MILLIMETERS MIN	MILLIMETERS MAX
A	0.160	0.180	4.06	4.57
B	0.130	0.150	3.30	3.81
C	0.075	0.095	1.90	2.41
D	0.077	0.083	1.96	2.11
H	0.0020	0.0060	0.051	0.152
J	0.006	0.012	0.15	0.30
K	0.030	0.050	0.76	1.27
P	0.020 REF		0.51 REF	
S	0.205	0.220	5.21	5.59

예제 2) SOT23

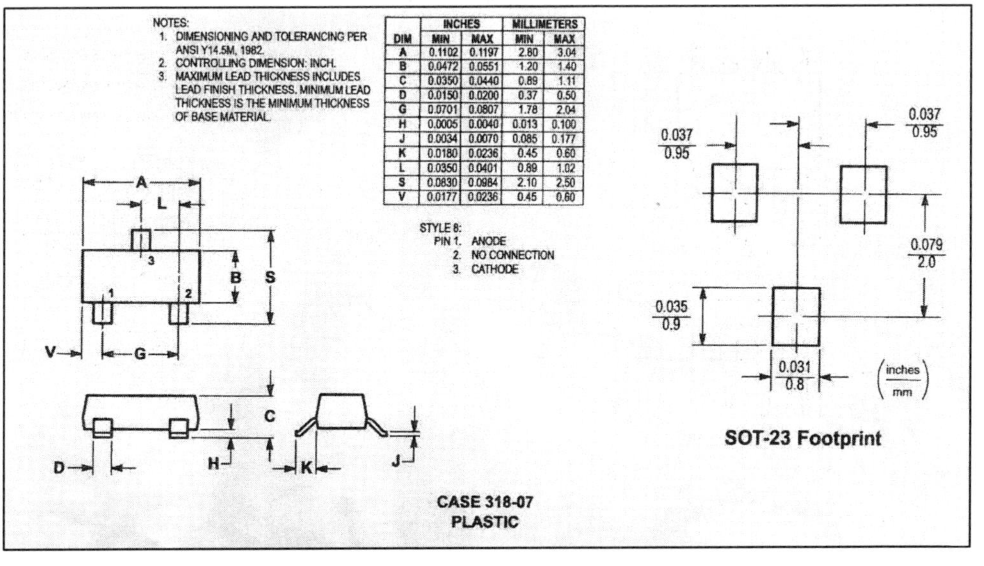

DIM	INCHES MIN	INCHES MAX	MILLIMETERS MIN	MILLIMETERS MAX
A	0.1102	0.1197	2.80	3.04
B	0.0472	0.0551	1.20	1.40
C	0.0350	0.0440	0.89	1.11
D	0.0150	0.0200	0.37	0.50
G	0.0701	0.0807	1.78	2.04
H	0.0005	0.0040	0.013	0.100
J	0.0034	0.0070	0.085	0.177
K	0.0180	0.0236	0.45	0.60
L	0.0350	0.0401	0.89	1.02
S	0.0830	0.0984	2.10	2.50
V	0.0177	0.0236	0.45	0.60

예제 3) 1열 Header

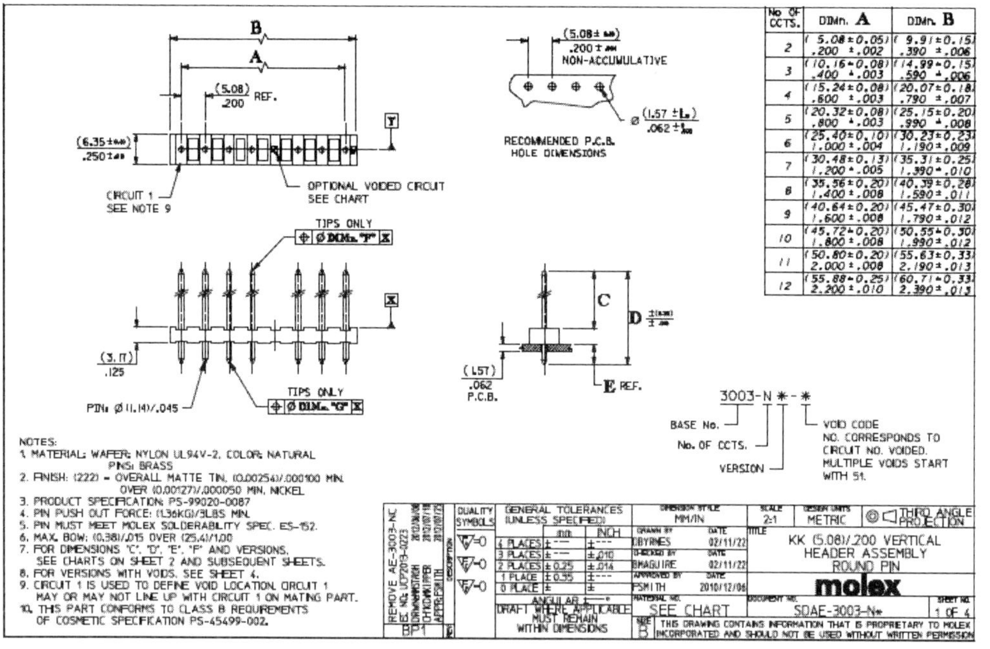

7. PAD 사이즈 결정법

예제 1) DIP8

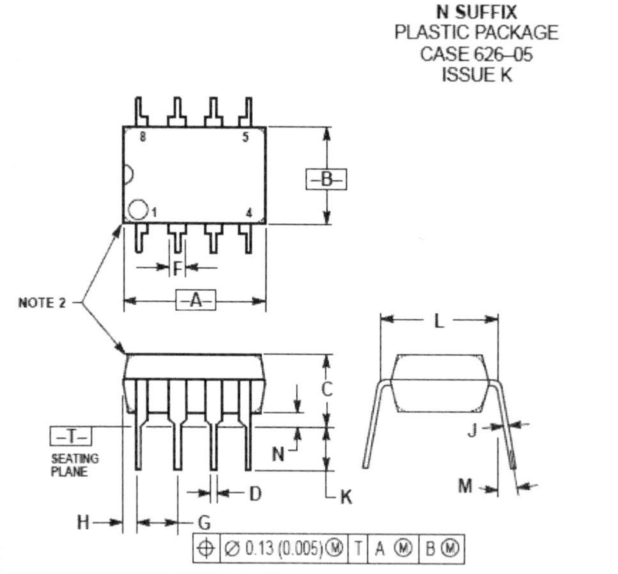

DIM	MILLIMETERS MIN	MILLIMETERS MAX	INCHES MIN	INCHES MAX
A	9.40	10.16	0.370	0.400
B	6.10	6.60	0.240	0.260
C	3.94	4.45	0.155	0.175
D	0.38	0.51	0.015	0.020
F	1.02	1.78	0.040	0.070
G	2.54 BSC		0.100 BSC	
H	0.76	1.27	0.030	0.050
J	0.20	0.30	0.008	0.012
K	2.92	3.43	0.115	0.135
L	7.62 BSC		0.300 BSC	
M	—	10°	—	10°
N	0.76	1.01	0.030	0.040

예제 2) SO8

D SUFFIX
PLASTIC PACKAGE
CASE 751–05
(SO–8)
ISSUE R

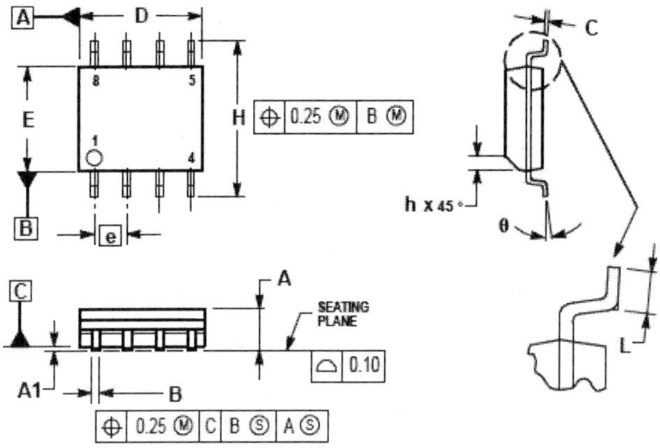

NOTES:
1. DIMENSIONING AND TOLERANCING PER ASME Y14.5M, 1994.
2. DIMENSIONS ARE IN MILLIMETERS.
3. DIMENSION D AND E DO NOT INCLUDE MOLD PROTRUSION.
4. MAXIMUM MOLD PROTRUSION 0.15 PER SIDE.
5. DIMENSION B DOES NOT INCLUDE MOLD PROTRUSION. ALLOWABLE DAMBAR PROTRUSION SHALL BE 0.127 TOTAL IN EXCESS OF THE B DIMENSION AT MAXIMUM MATERIAL CONDITION.

DIM	MILLIMETERS	
	MIN	MAX
A	1.35	1.75
A1	0.10	0.25
B	0.35	0.49
C	0.18	0.25
D	4.80	5.00
E	3.80	4.00
e	1.27 BSC	
H	5.80	6.20
h	0.25	0.50
L	0.40	1.25
θ	0°	7°

8. PCB Decal (PAD size 미표기)

예제 1) 1N914

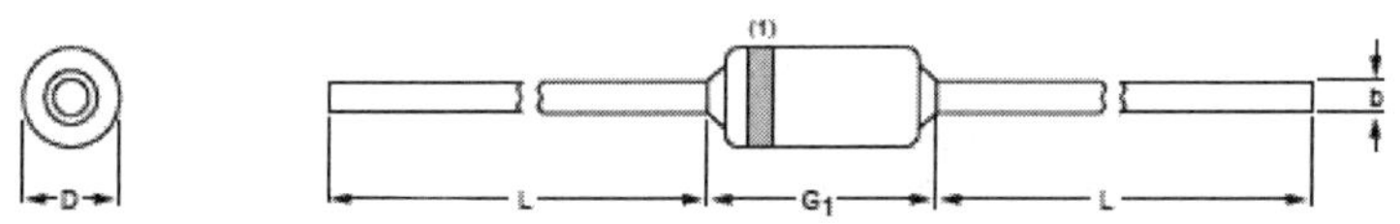

0 1 2 mm
scale

DIMENSIONS (mm are the original dimensions)

UNIT	b max.	D max.	G_1 max.	L min.
mm	0.56	1.85	4.25	25.4

Note

1. The marking band indicates the cathode.

OUTLINE VERSION	REFERENCES				EUROPEAN PROJECTION	ISSUE DATE
	IEC	JEDEC	EIAJ			
SOD27	A24	DO-35	SC-40			97-06-09

예제 2) LM741

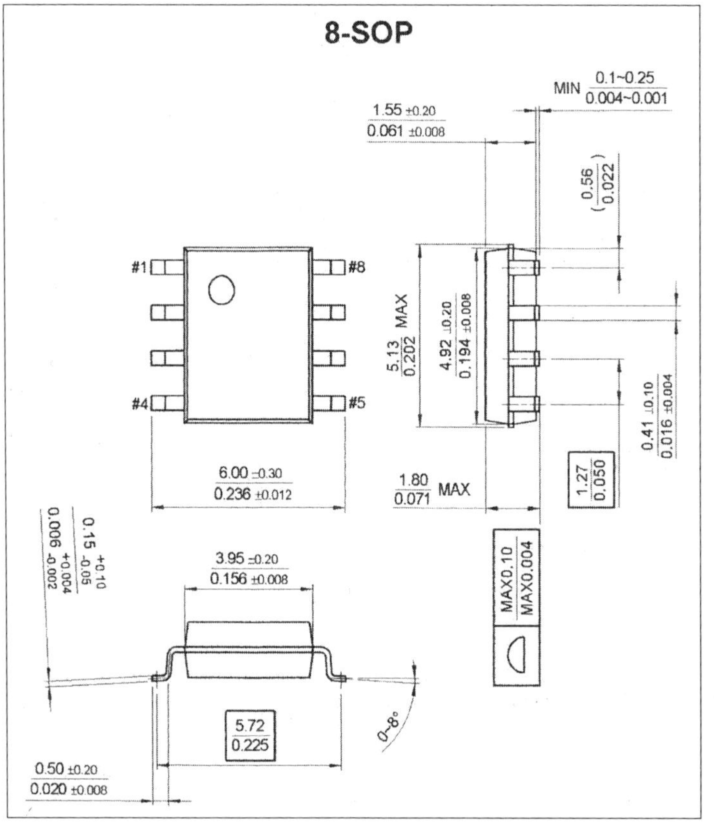

예제 3) SC70

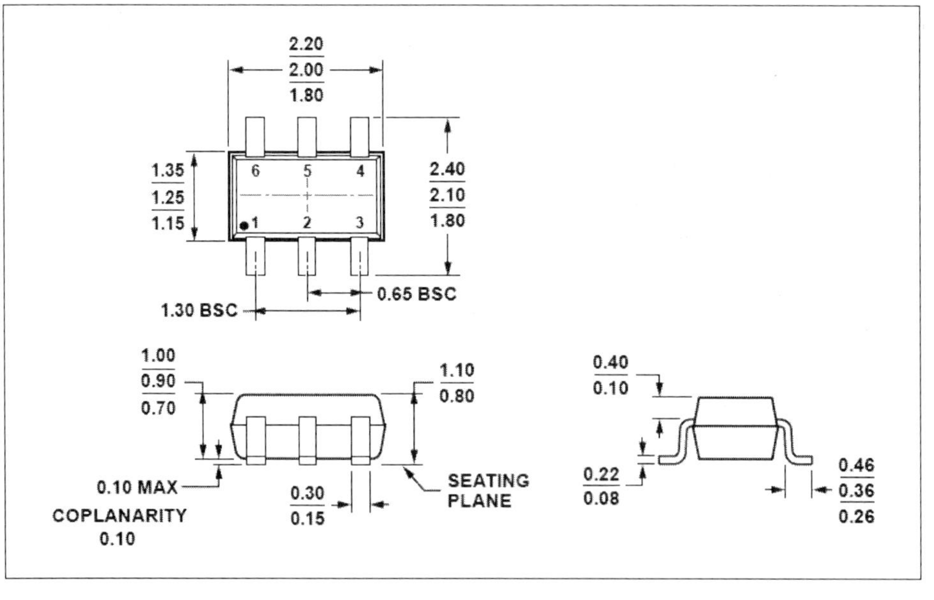

9. Board outline

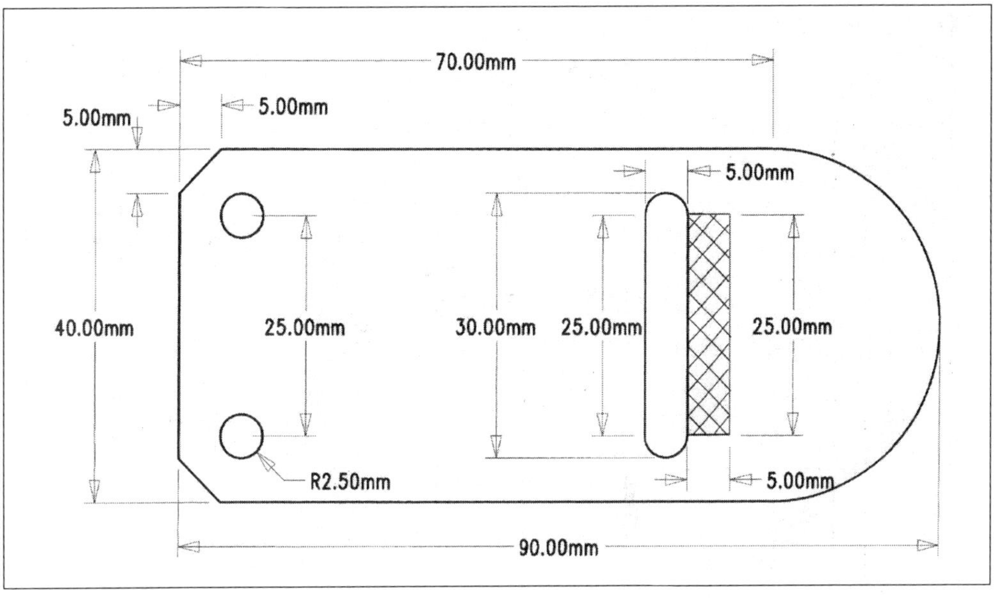

10. 실습회로 Board 그리기

PCB 가로 * 세로

11. 2진화 10진수 회로 설계하기

예제 1) 쉬운 회로도

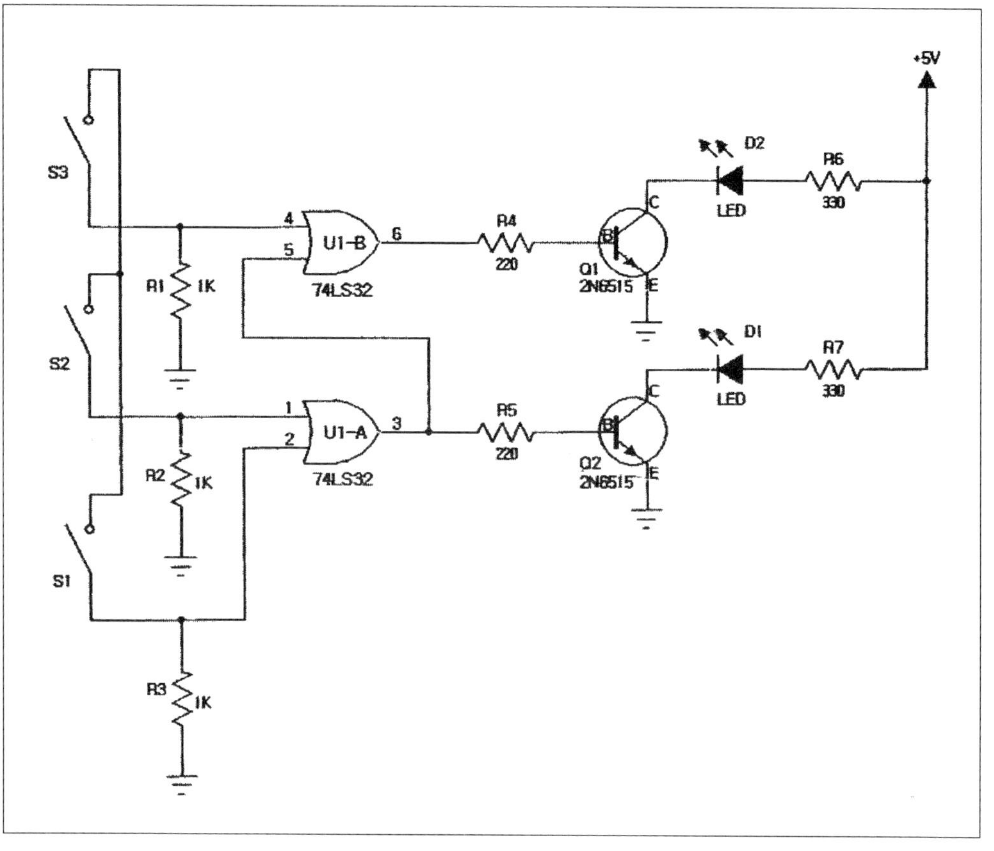

예제 2) 부품표

No.	수량	Ref-Des.	Part Type	PCB Decal	Value
1	2	Q1-2	2N6515 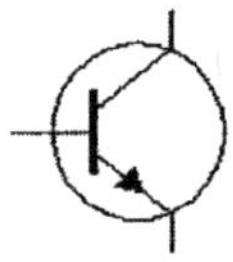	TO-220-UP 	
2	1	U1	74LS32 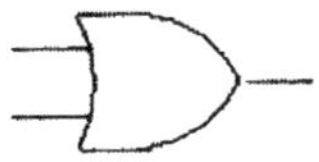	DIP14 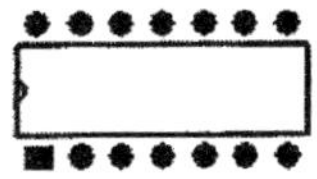	
3	2	D1-2	LED 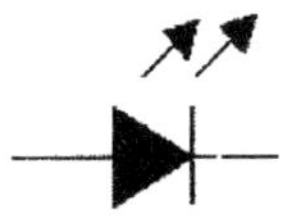	LED 	
4	3	S1-3	SW-NKK-JF15 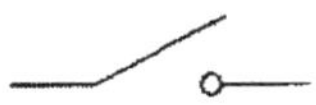	SW 	
5	3	R1-3	RES-1/8W 	R1/8W 	1K
6	2	R4-5	RES-1/8W	R1/8W	220
7	2	R6-7	RES-1/8W	R1/8W	330
Total	15				

예제 3) pcb size

50 × 60 mm

Upgrade

FB Lead type, With Conductive Polymer Series

Hi-CAP

- Low ESR, high ripple current
- Load life for 2000 hours at 105°C
- Complied to the RoHS directive

FB ⇒ **FJ**
Low ESR
High Cap.

Item	Characteristics		
Operating temperature range	-55 ~ +105°C		
Leakage current max.*	Less than or equal to the value of Table1		
Capacitance tolerance	±20% at 120Hz, 20°C		
Dissipation factor max.	Less than or equal to the value of Table1		
ESR	Less than or equal to the value of Table1		
Temperature characteristics (Impedance ratio at 100kHz)	Z-55°C / Z+20°C	Z+105°C / Z+20°C	
	0.75 ~ 1.25	0.75 ~ 1.25	
Load life (after application of the rated voltage for 2000 hours at 105°C, In case of 25WV is applied 20V)	Leakage current	Less than specified value	
	Capacitance change	Within ±20% of initial value	
	tanδ	Less than 150% of specified value	

* In case of some problems for measured values, measure the after applying rated voltage for 2.5 to 20V products or 20V derating voltage for 25V products for 120 minutes at 105°C.

DRAWING

Unit : mm

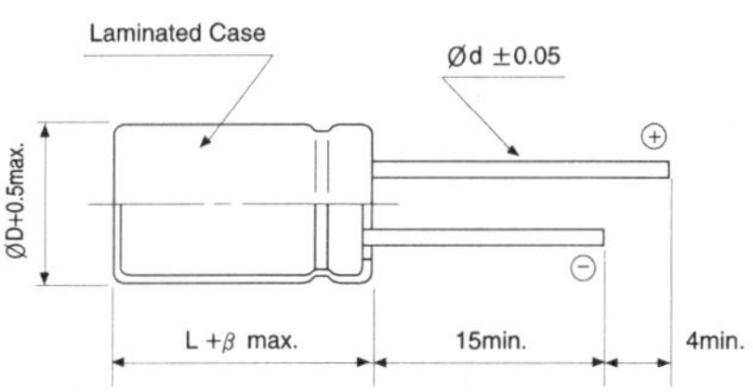

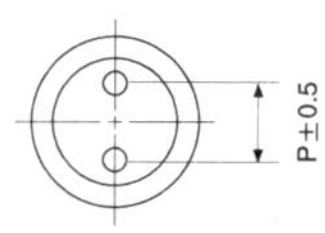

PART NUMBER SYSTEM (See Page 52)

(See Page 52)

Size	ØD	L	P	Ød	β
5×9	5	9.0	2.5	0.45	1.5
6.3×6	6.3	6.0	2.5	0.45	1.5
6.3×8	6.3	8.0	2.5	0.45	1.5
8×7	8.0	7.0	2.5	0.45	1.5
8×9	8.0	9.0	3.5	0.60	1.5
8×12	8.0	12.0	3.5	0.60	1.5
10×13	10.0	13.0	5.0	0.60	1.5

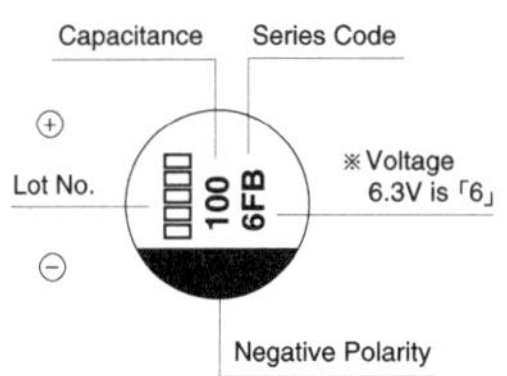

FB Series

● Table 1. FB(Lead type) Series Characteristics List

WV	μF	ØD(mm)	L(mm)	ESR(mΩ)max. 100~300kHz	Ripple current (mA rms)at 105°C 100kHz	Tangent of loss angle(max.)	Leakage Current (μA)(max.)
2.5	390	6.3	8	20	3160	0.10	195
2.5	560	8	9	7	6100	0.10	280
2.5	680	8	9	7	6100	0.10	340
2.5	820	8	9	7	6100	0.10	410
2.5	1000	8	9	7	6100	0.10	500
2.5	1500	8	12	8	5500	0.10	750
2.5	2700	10	13	8	5560	0.10	1350
2.5	3300	10	13	8	6650	0.10	1650
4	100	6.3	6	40	1810	0.10	80
4	150	6.3	6	40	1810	0.10	120
4	220	6.3	8	35	2560	0.10	176
4	270	6.3	8	20	3160	0.10	216
4	330	6.3	8	24	3300	0.10	264
4	390	6.3	8	24	3300	0.10	312
4	470	8	9	8	5700	0.10	376
4	560	8	9	7	6100	0.10	448
4	680	8	9	7	6100	0.10	544
4	1200	8	12	8	5700	0.10	960
6.3	82	6.3	6	45	1700	0.10	103
6.3	150	6.3	8	35	2560	0.10	189
6.3	220	6.3	8	20	3160	0.10	277
6.3	330	8	9	28	3190	0.10	416
6.3	470	8	9	8	5700	0.10	592
6.3	680	8	9	7	6640	0.10	857
6.3	820	8	12	7	6640	0.10	1033
6.3	1000	8	12	8	6100	0.10	1260
6.3	1200	10	13	10	5560	0.10	1520
6.3	1500	10	13	10	5560	0.10	1890
10	47	6.3	6	25	2820	0.10	94
10	56	6.3	6	25	2820	0.10	112
10	68	6.3	6	25	2820	0.10	136
10	100	6.3	8	25	2820	0.10	200
10	120	6.3	8	35	2560	0.10	240
10	150	6.3	8	25	2820	0.10	300
10	270	8	12	9	4710	0.10	540
10	330	8	12	9	4710	0.10	660
10	390	8	12	9	5650	0.10	780
10	470	8	12	8	5650	0.10	940
10	560	10	13	8	5650	0.10	1120
10	680	10	13	7	6100	0.10	1360

FB Series

● Table 1. FB(Lead type) Series Characteristics List

WV	μF	ØD(mm)	L(mm)	ESR(mΩ)max. 100~300kHz	Ripple current (mA rms)at 105°C 100kHz	Tangent of loss angle(max.)	Leakage Current (μA)(max.)
12	330	5	9	15	3160	0.10	100
16	39	6.3	6	50	1620	0.10	124
16	82	6.3	8	25	2120	0.10	262
16	100	6.3	8	25	2820	0.10	320
16	180	8	12	16	4360	0.10	576
16	270	8	12	11	5000	0.10	864
16	330	10	13	10	6100	0.10	1056
16	470	10	13	10	6100	0.10	1504
20	22	6.3	6	60	1450	0.12	88
20	47	6.3	8	25	2820	0.12	188
20	56	6.3	8	25	2820	0.12	224
20	68	6.3	8	25	2820	0.12	272
20	100	8	12	24	3320	0.12	400
20	150	10	13	20	4320	0.12	600
20	560	10	13	12	5400	0.12	2240
25	6.8	6.3	6	80	1200	0.12	85
25	10	6.3	8	60	1500	0.12	125
25	18	6.3	8	40	2230	0.12	225
25	22	8	12	40	2230	0.12	275
25	33	8	12	30	2980	0.12	413
25	56	8	12	30	3800	0.12	700
25	68	8	12	25	3320	0.12	850
25	82	8	12	24	3320	0.12	1025
25	100	8	12	20	4320	0.12	1250
35	10	8	7	60	1300	0.12	70
35	120	10	13	18	4400	0.12	840
35	150	10	13	16	4400	0.12	1050

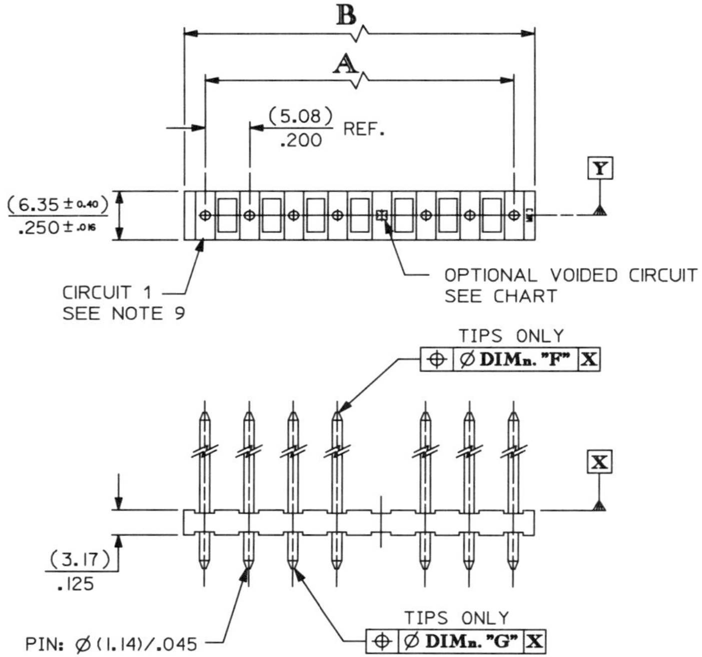

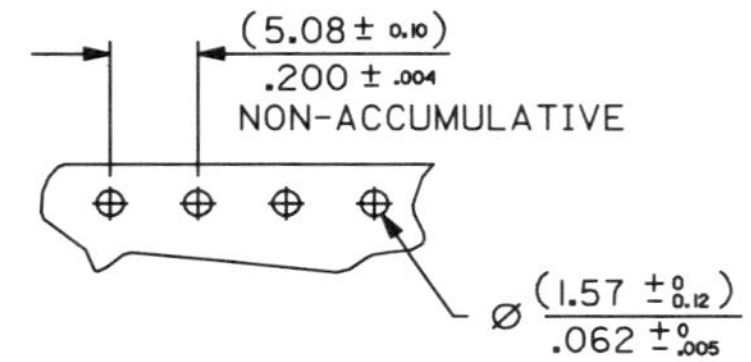

RECOMMENDED P.C.B.
HOLE DIMENSIONS

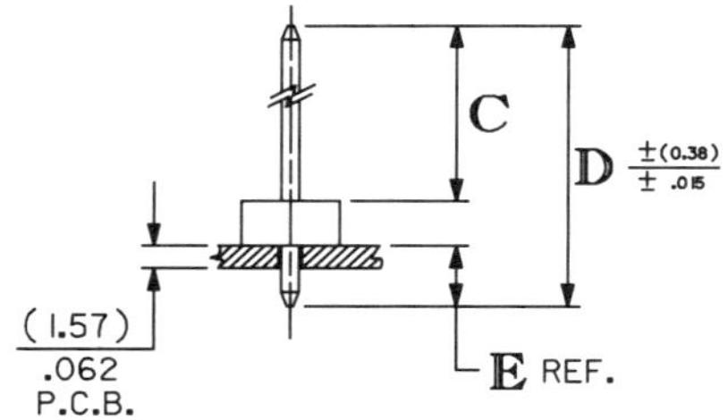

No OF CCTS.	DIMn. A	DIMn. B
2	(5.08 ± 0.05) .200 ± .002	(9.91 ± 0.15) .390 ± .006
3	(10.16 ± 0.08) .400 ± .003	(14.99 ± 0.15) .590 ± .006
4	(15.24 ± 0.08) .600 ± .003	(20.07 ± 0.18) .790 ± .007
5	(20.32 ± 0.08) .800 ± .003	(25.15 ± 0.20) .990 ± .008
6	(25.40 ± 0.10) 1.000 ± .004	(30.23 ± 0.23) 1.190 ± .009
7	(30.48 ± 0.13) 1.200 ± .005	(35.31 ± 0.25) 1.390 ± .010
8	(35.56 ± 0.20) 1.400 ± .008	(40.39 ± 0.28) 1.590 ± .011
9	(40.64 ± 0.20) 1.600 ± .008	(45.47 ± 0.30) 1.790 ± .012
10	(45.72 ± 0.20) 1.800 ± .008	(50.55 ± 0.30) 1.990 ± .012
11	(50.80 ± 0.20) 2.000 ± .008	(55.63 ± 0.33) 2.190 ± .013
12	(55.88 ± 0.25) 2.200 ± .010	(60.71 ± 0.33) 2.390 ± .013

3003-N * - *

- BASE No.
- No. OF CCTS.
- VERSION
- VOID CODE NO. CORRESPONDS TO CIRCUIT NO. VOIDED. MULTIPLE VOIDS START WITH 51.

NOTES:

1. MATERIAL: WAFER: NYLON UL94V-2, COLOR: NATURAL
   PINS: BRASS
2. FINISH: (222) = OVERALL MATTE TIN, (0.00254)/.000100 MIN. OVER (0.00127)/.000050 MIN. NICKEL
3. PRODUCT SPECIFICATION: PS-99020-0087
4. PIN PUSH OUT FORCE: (1.36KG)/3LBS MIN.
5. PIN MUST MEET MOLEX SOLDERABILITY SPEC. ES-152.
6. MAX. BOW: (0.38)/.015 OVER (25.4)/1.00
7. FOR DIMENSIONS "C", "D", "E", "F" AND VERSIONS, SEE CHARTS ON SHEET 2 AND SUBSEQUENT SHEETS.
8. FOR VERSIONS WITH VOIDS, SEE SHEET 4.
9. CIRCUIT 1 IS USED TO DEFINE VOID LOCATION. CIRCUIT 1 MAY OR MAY NOT LINE UP WITH CIRCUIT 1 ON MATING PART.
10. THIS PART CONFORMS TO CLASS B REQUIREMENTS OF COSMETIC SPECIFICATION PS-45499-002.

REV	DESCRIPTION	
BP1	REMOVE AE-3003-NC EC NO: UCP2013-0223 DRWN:MMSTROH CH'KD:MKIPPER APPR:FSMITH	 2012/06/08 2012/07/18 2012/07/25

QUALITY SYMBOLS: F_A=0, F_C=0, F_P=0

GENERAL TOLERANCES (UNLESS SPECIFIED)	mm	INCH
4 PLACES	± ---	± ---
3 PLACES	± ---	± .010
2 PLACES	± 0.25	± .014
1 PLACE	± 0.35	± ---
0 PLACE	±	±
ANGULAR ± ---°		

DRAFT WHERE APPLICABLE MUST REMAIN WITHIN DIMENSIONS

DIMENSION STYLE: MM/IN
SCALE: 2:1
DESIGN UNITS: METRIC
THIRD ANGLE PROJECTION

		DATE
DRAWN BY	DBYRNES	02/11/22
CHECKED BY	BMAGUIRE	02/11/22
APPROVED BY	FSMITH	2010/12/06

TITLE: KK (5.08)/.200 VERTICAL HEADER ASSEMBLY ROUND PIN

molex

MATERIAL NO.: SEE CHART
DOCUMENT NO.: SDAE-3003-N*
SHEET NO.: 1 OF 4
SIZE: B

Tact Switch
YST-1102 Dip Type

Feature
6mm x 6mm outer dimension suites directly mounted on PC board.

Available operating forces : 160gf, 250gf

Applications
For various operation in audio equipment, office devices, communication devices, measuring devices, TVs and VCRs.

Specifications

Item	Rating	Travel	Contact Registance	Bounce	Operating Force	Operating Life
Value	DC 12V 50mA MAX	0.25 ± 0.1mm	100mΩ MAX	10m sec MAX	160 ± 50gf	80,000 Cycle

Dimensions

Dimensions

Height (L)

TAG	(Blank)	A	B	C	D
Value (mm)	4.3	5.0	6.0	6.3	7.0
TAG	E	F	G	H	I
Value (mm)	7.3	8.0	8.5	9.5	12.5
TAG	J	K	L	M	N
Value (mm)	13.0	13.5	17.0	9.0	11.0

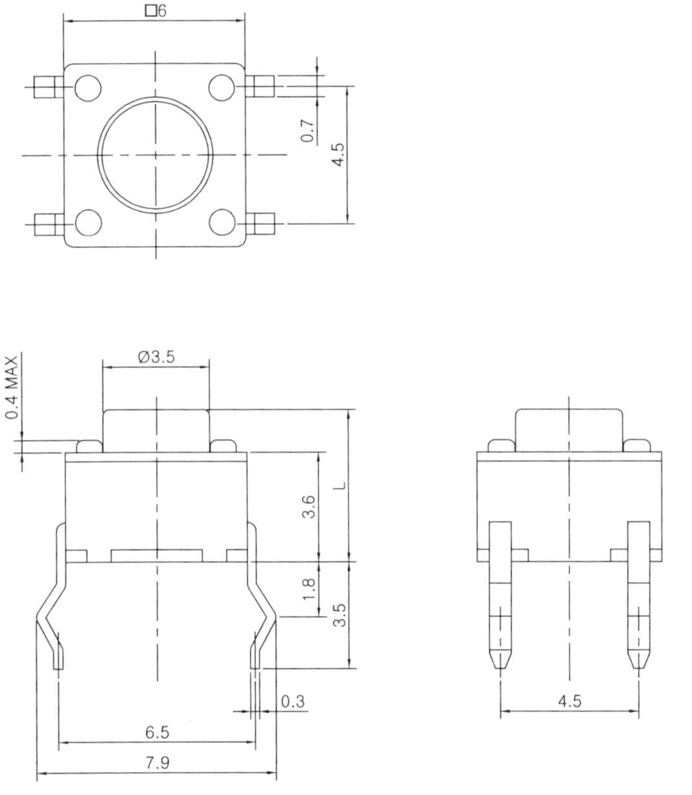

CIRCUIT DIAGRAM

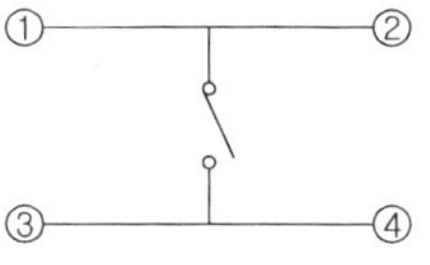

PCB MOUNTING
HOLE DIMENSIONS

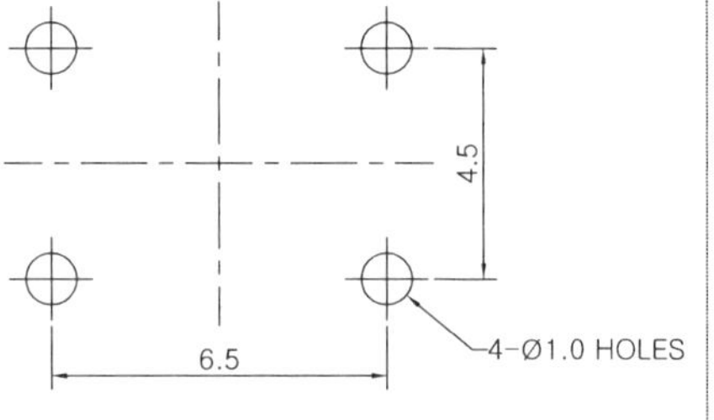

WTE
POWER SEMICONDUCTORS

1N4148 / LL4148

FAST SWITCHING DIODE

Features

- Fast Switching Speed
- Glass Package Version for High Reliability
- High Conductance
- Available in Both Through-Hole and Surface Mount Versions

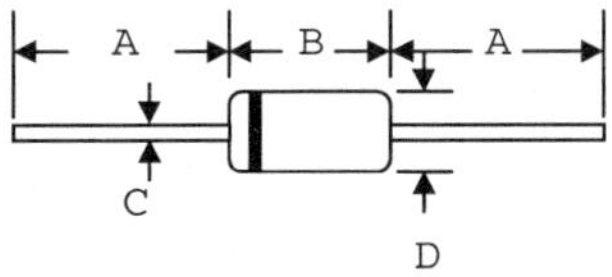

1N4148

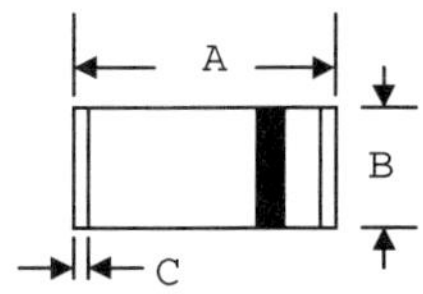

LL4148

Mechanical Data

- Case: DO-35, MiniMELF
- Terminals: Plated Leads Solderable per MIL-STD-202, Method 208
- Polarity: Cathode Band
- Weight: DO-35 0.13 grams
  MiniMELF 0.05 grams
- Marking: Cathode Band Only

DO-35		
Dim	Min	Max
A	25.40	—
B	—	4.00
C	—	0.60
D	—	2.00
All Dimensions in mm		

MiniMELF		
Dim	Min	Max
A	3.30	3.70
B	1.30	1.60
C	0.28	0.50
All Dimensions in mm		

Maximum Ratings @ T_A=25°C unless otherwise specified

Characteristic	Symbol	Value	Unit
Non-Repetitive Peak Reverse Voltage	V_{RM}	100	V
Peak Repetitive Reverse Voltage Working Peak Reverse Voltage DC Blocking Voltage	V_{RRM} V_{RWM} V_R	75	V
RMS Reverse Voltage	$V_{R(RMS)}$	53	V
Forward Continuous Current (Note 1)	I_{FM}	300	mA
Rectified Current (Average), Half Wave Rectification with Resistive Load and f ≧ 50MHz (Note 1)	I_O	150	mA
Non-Repetitive Peak Forward Surge Current @ t = 1.0s @ t = 1.0μs	I_{FSM}	1.0 2.0	A
Power Dissipation (Note 1) Derate Above 25°C	Pd	500 1.68	mW mW/°C
Thermal Resistance, Junction to Ambient Air (Note 1)	$R_{\theta JA}$	300	K/W
Operating and Storage Temperature Range	T_j, T_{STG}	-65 to +175	°C

ORDERING INFORMATION

Product No.◆	Package Type	Shipping Quantity
1N4148-T3	DO-35	10000/Tape & Reel
1N4148-TB	DO-35	5000/Tape & Box
1N4148	DO-35	1000 Units/Box
LL4148-T1	MiniMELF	2500/Tape & Reel
LL4148-T3	MiniMELF	10000/Tape & Reel

Products listed in bold are WTE Preferred devices.
◆T1 suffix refers to a 7" reel. T3 suffix refers to a 13" reel. TB suffix refers to Ammo Pack.
Shipping quantity given is for minimum packing quantity only. For minimum order quantity, please consult the Sales Department.

RECOMMENDED FOOTPRINT (MiniMELF)

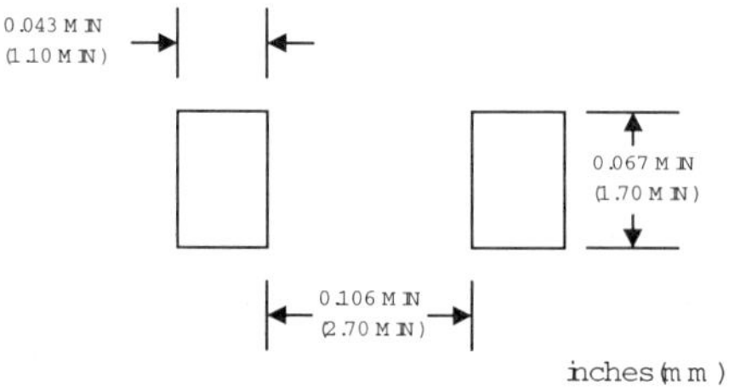

inches (mm)

Won-Top Electronics Co., Ltd.
No. 44 Yu Kang North 3rd Road, Chine Chen Dist., Kaohsiung, Taiwan
Phone: 886-7-822-5408 or 886-7-822-5410
Fax: 886-7-822-5417
Email: sales@wontop.com
Internet: http://www.wontop.com

We power your everyday.

SX-32

SX-22

CERAMIC SMD TYPE

CRYSTAL UNIT

SX-32, SX-22

- Four pad ceramic suface mount package
- AT cut
- Tight tolerance/stability
- Using highly reliable seam weld method
- Taping and reel available, automatic mountable
- Pb-free and RoHS compliance

■ ELECTRICAL SPECIFICATIONS

Frequency Range	12.000 to 50.000MHz
Operating Temperature Range	-10 to +60℃(STD), -20 to +70℃, -40 to 85℃, or specify
Storage Temperature Range	-40 to +85℃
Frequency Tolerance (at 25℃)	±15, ±30, ±50ppm(STD), or specify
Frequency stability Over Operating Temperature Range	See Table1 in page 4
Load Capacitance(CL)	12pF(STD), Custom CL≥8pF or Series Resonant
Drive Level	10µW(100µW max.)
Shunt Capacitance(Co)	5.0pF max.
Aging (at 25℃)	±3.0ppm/year max.
Insulation Resistance	500MΩ min. at 100V$_{DC}$

■ EQUIPVALENT SERIES RESISTANCE(ESR)

Frequency Range(MHz)	ESR(Ω MAX.)	Mode of Oscillation	Package
12.000 ~ 13.999	150	Fundmental	SX-32
14.000 ~ 15.999	100	Fundmental	SX-32
16.000 ~ 19.999	80 / 120	Fundmental	SX-32 / SX-22
20.000 ~ 29.999	60 / 80	Fundmental	SX-32 / SX-22
30.000 ~ 54.000	50 / 50	Fundmental	SX-32 / SX-22

■ MECHANICAL DIMENSIONS

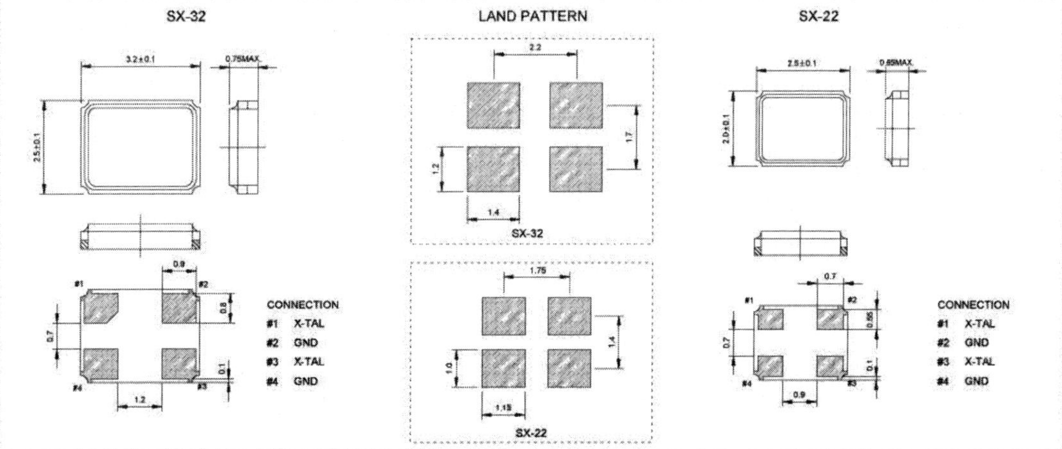

CDMC10D50/T150

sumida

PROVISIONAL

RoHS

Features

- Metal compound molding type construction.
- Magnetically shielded.
- Qualified AEC-Q200
- Suitable for Large Current
- Operating temperature range: - 40deg C～+150deg C (including coil's self temperature rise)

Applications

- LED Head light for Automobile
- ECU for Airbag, Water and Oil pump etc
- DC/DC converter

Dimension - [mm]

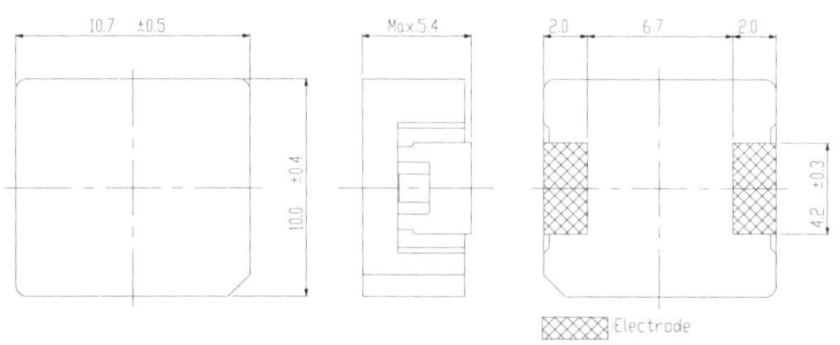

Reference Land pattern - [mm]

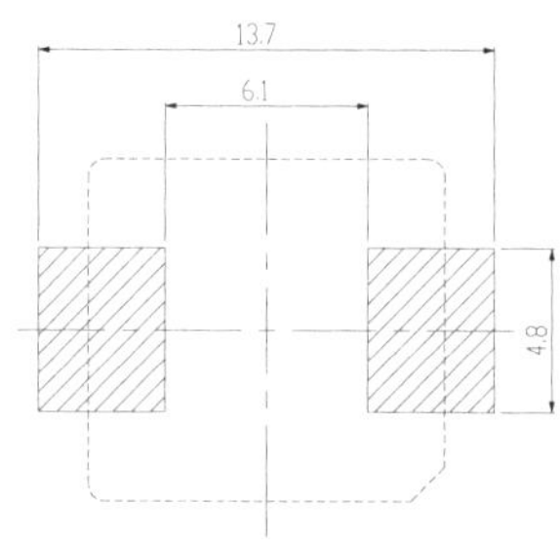

Electrical Characteristics

Part No.	Inductance (μH) (±20%) ∗1	D.C.R. (mΩ) at 20degC (±10%)	Saturation Current (A) ∗2			Temperature Rise Current (A) ∗3
			at 20degC	at 110degC	at 150degC	
CDMC10D50T150NP-1R1MC	1.1	3.3	32.0	30.0	30.0	19.1
CDMC10D50T150NP-1R6MC	1.6	4.1	26.8	24.5	24.3	16.9
CDMC10D50T150NP-2R2MC	2.2	5.1	24.4	22.4	22.2	15.9
CDMC10D50T150NP-2R7MC	2.7	6.0	21.0	19.6	19.2	14.5
CDMC10D50T150NP-3R3MC	3.3	6.8	18.6	17.3	17.0	14.0
CDMC10D50T150NP-4R3MC	4.3	10.0	15.9	14.7	14.6	10.7

∗1 Inductance Measuring frequency: 100kHz, 0.1V

∗2 Saturation current: DC current which becomes inductance value drop down to 30% from the initial value.

∗3 Temperature rise current : DC current which becomes △T=40 deg C (Ta=20 deg C) in component body temperature.
Board conditions: FR4, Copper=70μm, four-layer PWB, t=1.6mm

∗In order to select the proper inductor for your circuit design, it should be checked the inductor rated current for heat generation in the worst case. Under normal operational condition, it shall not exceed the maximum operating temperature (150 deg C.)

∗Allowable current of inductor is either saturation current or temperature rise current, whichever is lower.

∗Absolute maximum voltage 40V DC.

∗Take the static electricity measures to prevent deterioration of electric characteristics.

Note: This datasheet might be slightly modified without notice in order to improve the quality and reliability of the product. Please contact an appropriate our sales in your regions for checking final spec before you place an order.

Package Details
SOT-23 Case

Central Semiconductor Corp.

Mechanical Drawing

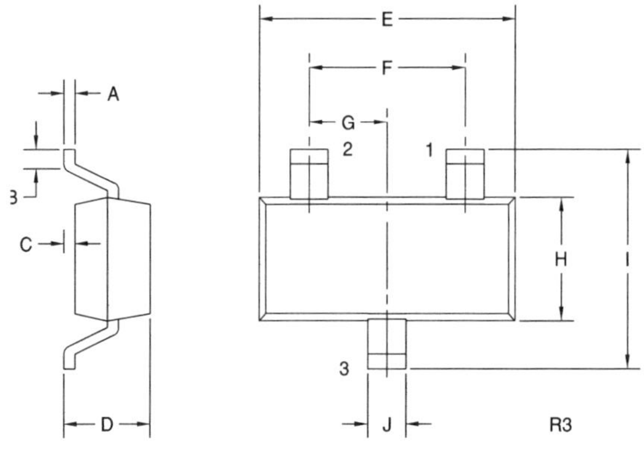

DIMENSIONS				
	INCHES		MILLIMETERS	
SYMBOL	MIN	MAX	MIN	MAX
A	0.003	0.007	0.08	0.18
B	0.006	-	0.15	-
C	-	0.005	-	0.13
D	0.035	0.043	0.89	1.09
E	0.110	0.120	2.80	3.05
F	0.075		1.90	
G	0.037		0.95	
H	0.047	0.055	1.19	1.40
I	0.083	0.098	2.10	2.49
J	0.014	0.020	0.35	0.50

SOT-23 (REV: R3)

Lead Code:
Reference individual device datasheet.

Part Marking: 2-4 Character Alpha/Numeric Code

Mounting Pad Geometry (Dimensions in mm)

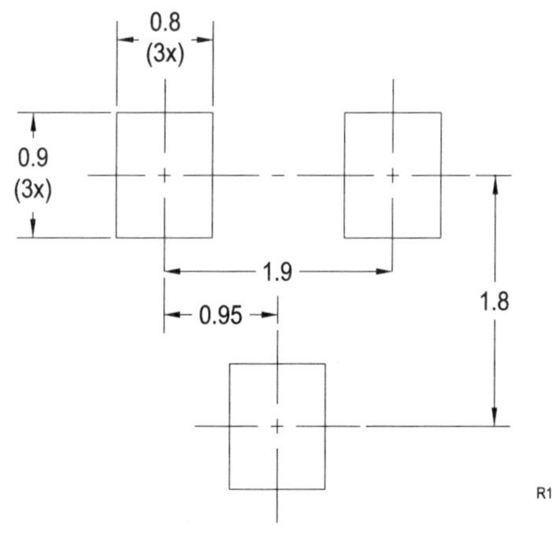

R3 (4-March 2010)

MODEL	SPECIFICATION DIP SWITCH	SP-003	
DI series IC Type	RoHS&REACH Compliant / Halogen Free		

DIJ

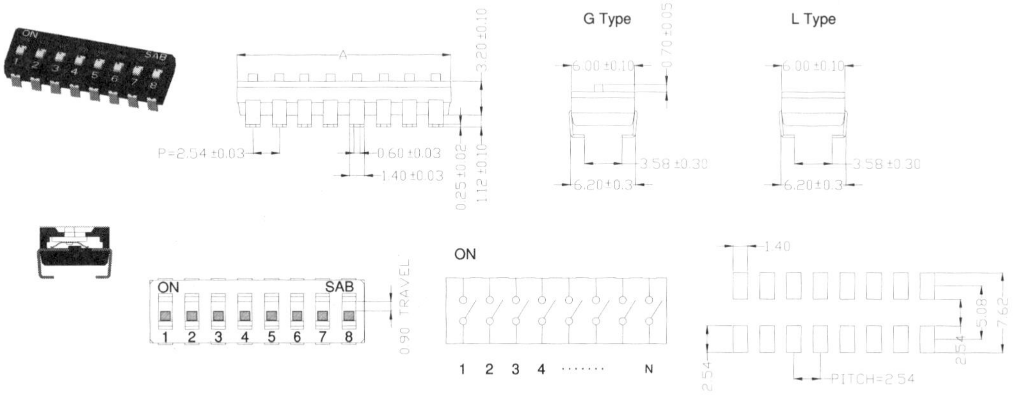

No. Pos.	Dim. "A"
1	2.98
2	5.52
3	8.06
4	10.60
5	13.14
6	15.68
7	18.22
8	20.76
9	23.30
10	25.84
12	30.92

DIS

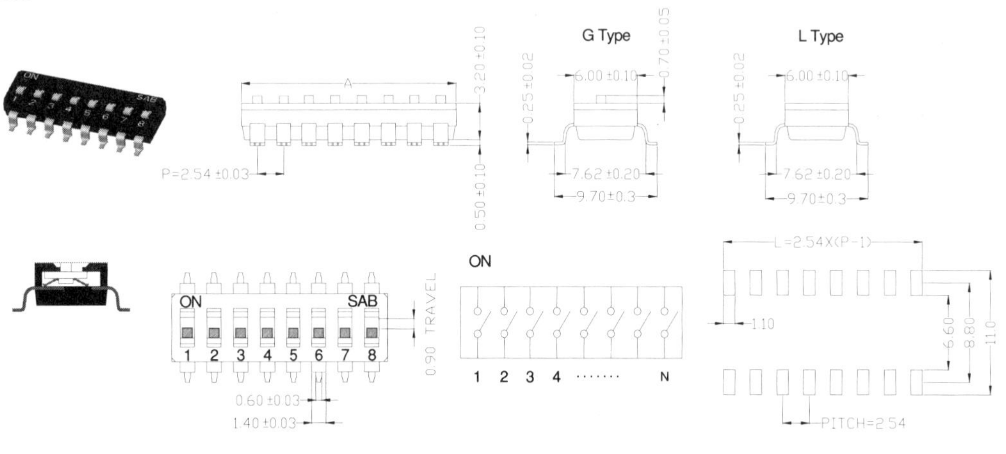

No. Pos.	Dim. "A"
1	2.98
2	5.52
3	8.06
4	10.60
5	13.14
6	15.68
7	18.22
8	20.76
9	23.30
10	25.84
12	30.92

DIA-H

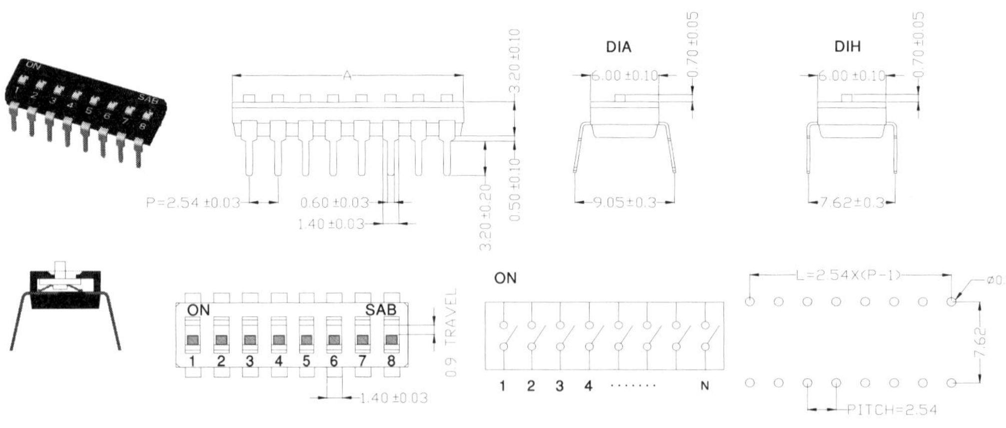

No. Pos.	Dim. "A"
1	2.98
2	5.52
3	8.06
4	10.60
5	13.14
6	15.68
7	18.22
8	20.76
9	23.30
10	25.84
12	30.92

MODEL	SPECIFICATION DIP SWITCH RoHS&REACH Compliant / Halogen Free	SP-004	
DI series IC Type			

Tube Package

POLE	1	2	3	4	5	6	7	8	9	10	12
TUBE	174	95	65	48	39	33	28	25	22	20	17
BOX	17400	9500	6500	4800	3900	3300	2800	2500	2200	2000	1700
CTN 1	34800	19000	13000	9600	7800	6600	5600	5000	4400	4000	3400
CTN 2	69600	38000	26000	19200	15600	13200	11200	10000	8800	8000	6800

Tape & Reel Package

◆ **DIS**

POLE	01	02	03	04	05	06	07	08	09	10	12
Pitch (mm)	8	8	16	16	16	16	16	16	16	16	16
Width (mm)	16	16	16	24	24	24	32	32	44	44	44
G type / Reel	1500	1500	800	800	800	800	800	800	800	800	800
L type / Reel	2000	2000	1000	1000	1000	1000	1000	1000	1000	1000	1000

◆ **DIJ**

POLE	01	02	03	04	05	06	07	08	09	10	12
Pitch (mm)	8	12	12	12	12	12	12	12	12	12	12
Width (mm)	16	16	16	24	24	24	32	32	44	44	44
G type / Reel	1500	900	900	900	900	900	900	900	900	900	900
L type / Reel	1600	1100	1100	1100	1100	1100	1100	1100	1100	1100	1100

DIS

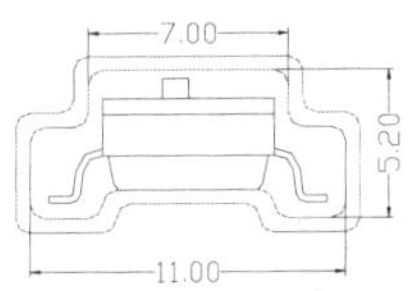

DIA/H

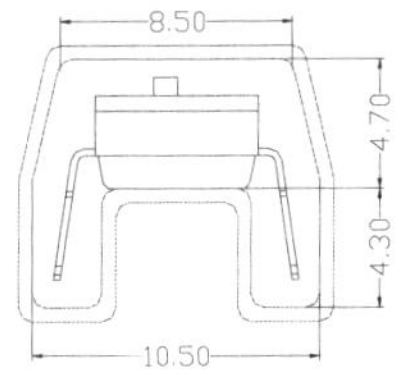

DIJ01/DIJ02

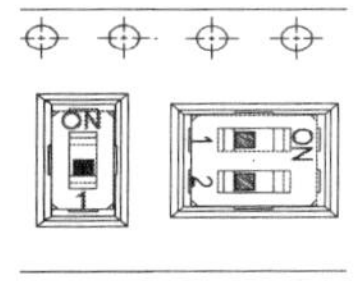

DIS01/DIS02

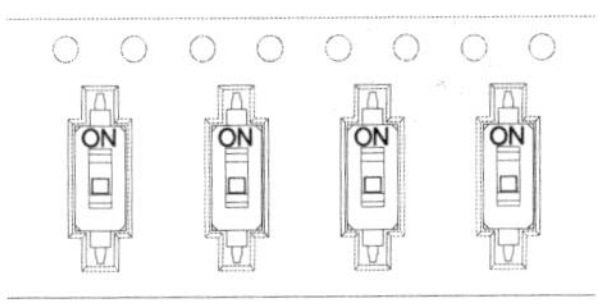

DIS03~12

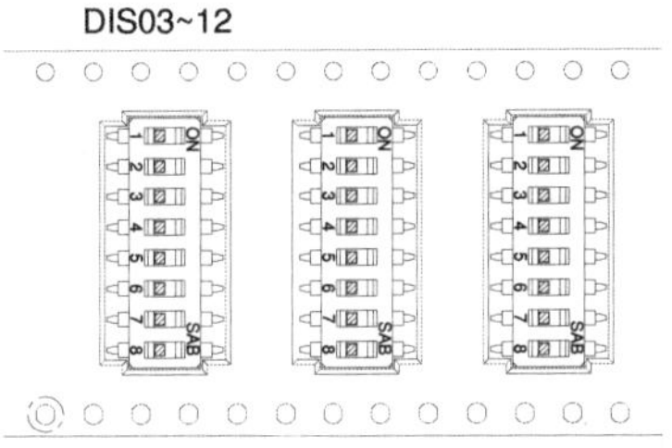

DIJ03~12

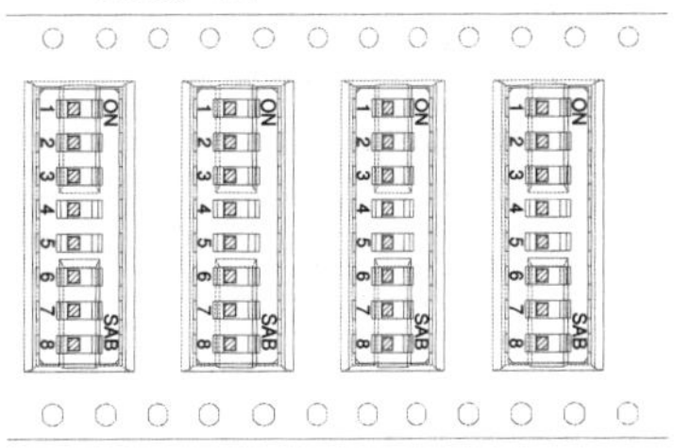

Box Dimension

◆ Tube .... 57.0cm x 13.0cm x 12.5cm

◆ Tape & Reel .... 34.5cm x 34.5cm x 15.0cm

Carton Dimension

◆ Tube
2 Boxes........................ 58.5cm x 29.0cm x 15.0cm
4 Boxes........................ 58.5cm x 28.0cm x 27.0cm

◆ Tape & Reel
1 Box........................... 34.5cm x 34.5cm x 15.0cm
2 Boxes........................ 36.0cm x 27.5cm x 37.0cm

USB CONNECTOR
TYPE B SINGLE PORT & TYPE B PLUG
USB SERIES

USB-B-S-RA

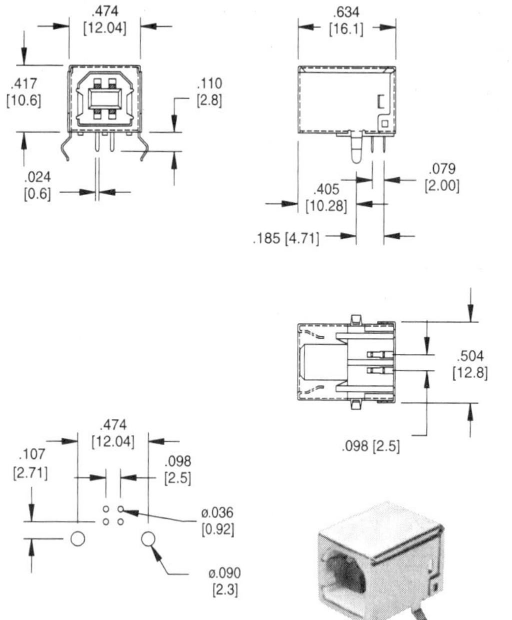

Recommended PCB Layout

USB-B-S-RA-TSMT

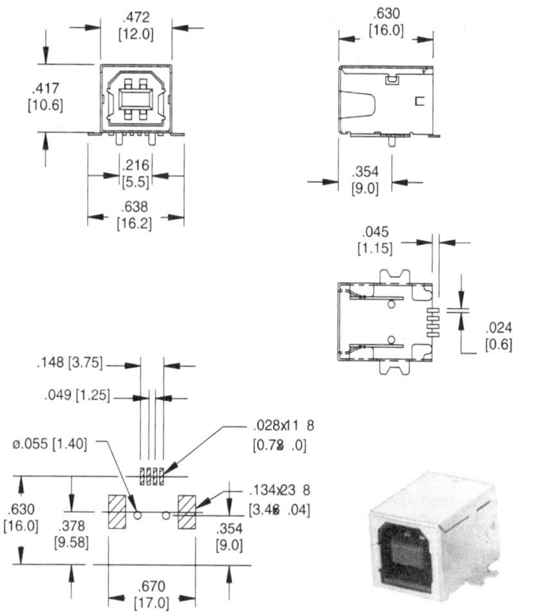

Recommended PCB Layout

USB-B-S-VT

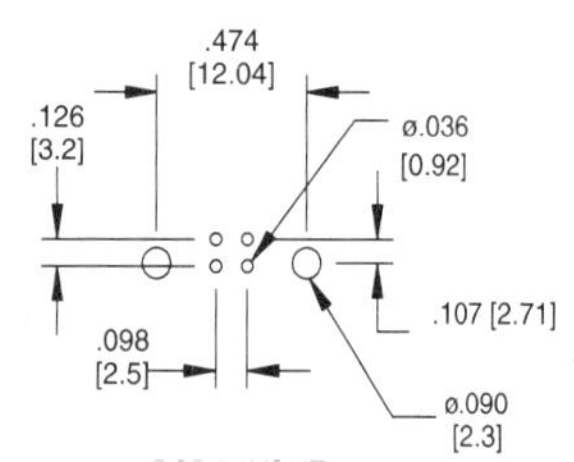

Recommended PCB Layout
(Component Side)

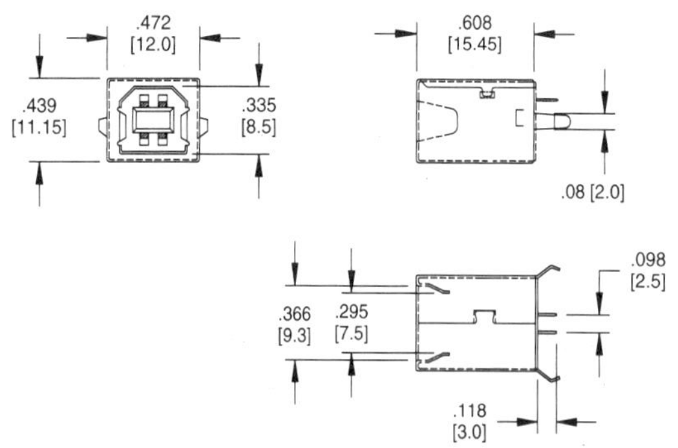

USB-BP-S

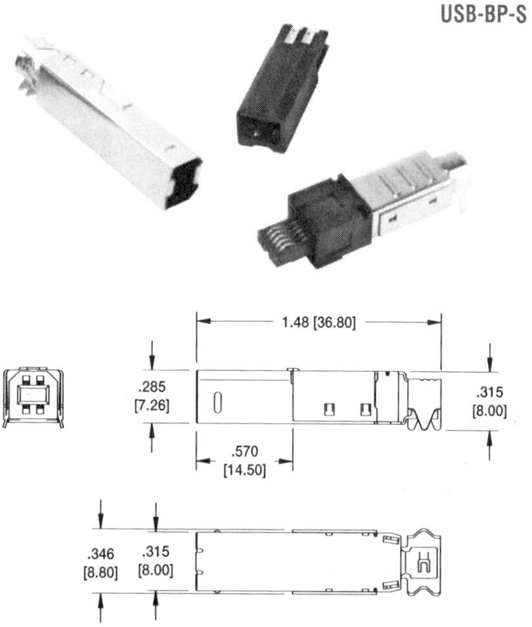

Honglitronic

ATTENTION
OBSERVE PRECAUTIONS
FOR HANDLING
ELECTROSTATIC
DISCHARGE
SENSITIVE
DEVICES

HL-A-3528H196W-A170 WHITE

Features

- SINGLE COLOR.
- SUITABLE FOR ALL SMT ASSEMBLY AND SOLDER PROCESS.
- AVAILABLE ON TAPE AND REEL.
- IDEAL FOR BACKLIGHTING.
- PACKAGE :2000PCS /REEL.

Description

The White source color devices are made with Gallium Arsenide Phosphide on Gallium Phosphide White Light Emitting Diode.

Package Dimensions

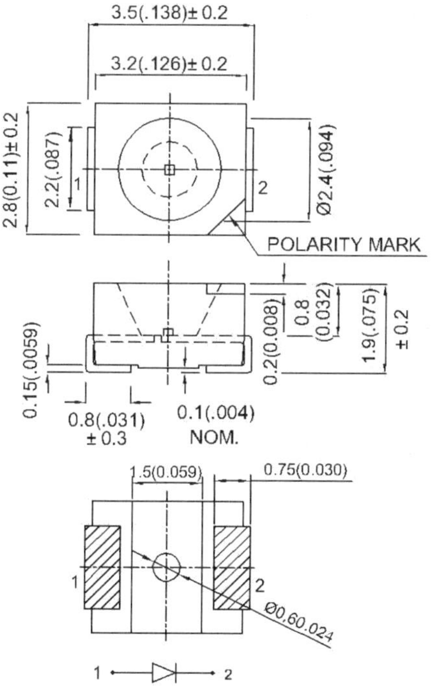

Notes:

1. All dimension units are millimeters.
2. All dimension tolerance is ±0.2mm unless otherwise noted.
3. An epoxy meniscus may extend about 1.5mm down the leads.
4. Burr around bottom of epoxy may be 0.5mm max..

SPEC NO : A1539 REV NO : V1 DATE : AVG/16/2006

APPROVED : CHECKED : DRAWN :

Honglitronic

Selection Guide

Part No.	Dice	Lens Type	Iv (mcd) @ 20mA		Viewing Angle
			Min.	Typ.	2θ1/2
HL-A-3528H196W-A170	WHITE (InGaN)	WATER CLEAR	750	1150	120°

Note:
1. θ1/2 is the angle from optical centerline where the luminous intensity is 1/2 the optical centerline value.

Electrical/Optical Characteristics at T_A=25℃

Symbol	Parameter	Device	Min.	Typ.	Units	Test Conditions
V_F	Forward Voltage	White	3.0	3.2	V	I_F=20mA
I_R	Reverse Current	White		5	uA	V_R = 5V
X	Chromaticity Coordinates	White		0.29		
Y				0.30		
C	Capacitance	White		100	pF	V_F=0V;f=1MHz

Absolute Maximum Ratings at T_A=25℃

Parameter	White	Units
Power dissipation	114	mW
DC Forward Current	30	mA
Peak Forward Current [1]	150	mA
Reverse Voltage	5	V
Operating/Storage Temperature	-40℃ To +85℃	

Note:
1. 1/10 Duty Cycle, 0.1ms Pulse Width.

HL-A-3528H196W-A170

SMT Reflow Soldering Instructions

Number of reflow process shall be less than 2 times and cooling process to normal temperature is required between first and second soldering process.

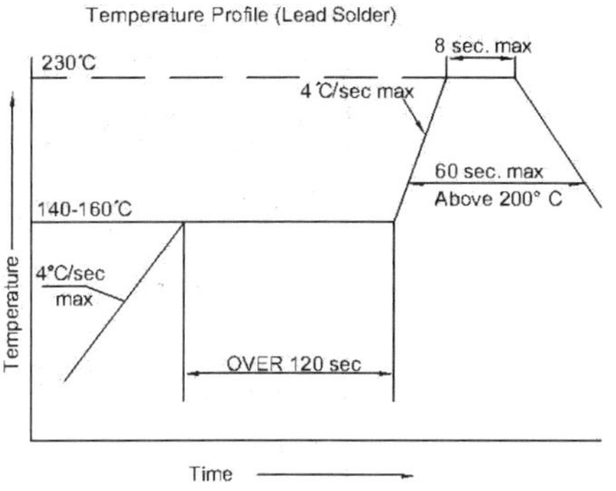

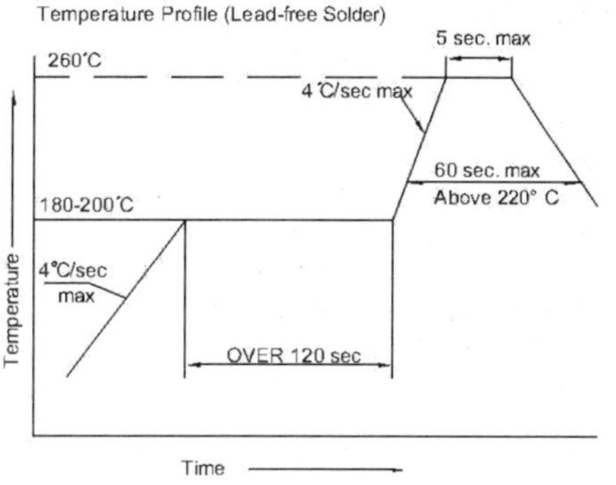

Recommended Soldering Pattern
(Units : mm)

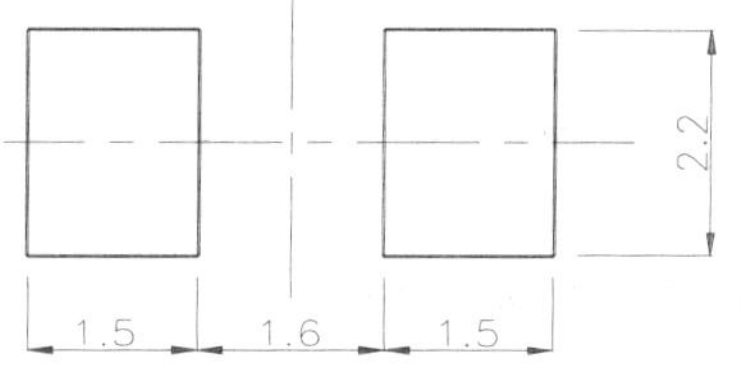

Tape Specifications
(Units : mm)

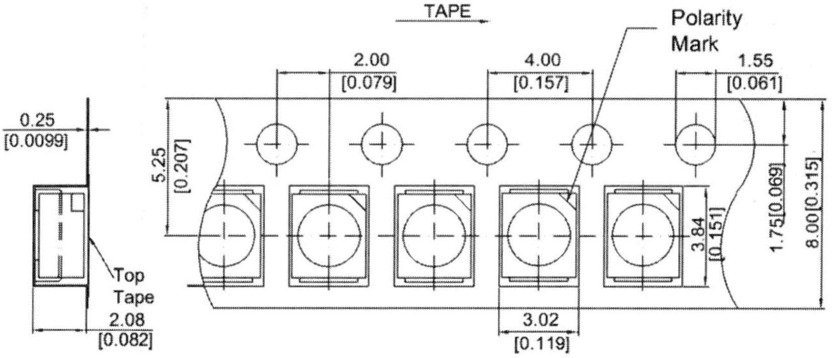

Features

- High-performance, Low-power Atmel®AVR® 8-bit Microcontroller
- Advanced RISC Architecture
  - 130 Powerful Instructions – Most Single-clock Cycle Execution
  - 32 × 8 General Purpose Working Registers
  - Fully Static Operation
  - Up to 16MIPS Throughput at 16MHz
  - On-chip 2-cycle Multiplier
- High Endurance Non-volatile Memory segments
  - 8Kbytes of In-System Self-programmable Flash program memory
  - 512Bytes EEPROM
  - 1Kbyte Internal SRAM
  - Write/Erase Cycles: 10,000 Flash/100,000 EEPROM
  - Data retention: 20 years at 85°C/100 years at 25°C[(1)]
  - Optional Boot Code Section with Independent Lock Bits
    In-System Programming by On-chip Boot Program
    True Read-While-Write Operation
  - Programming Lock for Software Security
- Peripheral Features
  - Two 8-bit Timer/Counters with Separate Prescaler, one Compare Mode
  - One 16-bit Timer/Counter with Separate Prescaler, Compare Mode, and Capture Mode
  - Real Time Counter with Separate Oscillator
  - Three PWM Channels
  - 8-channel ADC in TQFP and QFN/MLF package
    Eight Channels 10-bit Accuracy
  - 6-channel ADC in PDIP package
    Six Channels 10-bit Accuracy
  - Byte-oriented Two-wire Serial Interface
  - Programmable Serial USART
  - Master/Slave SPI Serial Interface
  - Programmable Watchdog Timer with Separate On-chip Oscillator
  - On-chip Analog Comparator
- Special Microcontroller Features
  - Power-on Reset and Programmable Brown-out Detection
  - Internal Calibrated RC Oscillator
  - External and Internal Interrupt Sources
  - Five Sleep Modes: Idle, ADC Noise Reduction, Power-save, Power-down, and Standby
- I/O and Packages
  - 23 Programmable I/O Lines
  - 28-lead PDIP, 32-lead TQFP, and 32-pad QFN/MLF
- Operating Voltages
  - 2.7V - 5.5V (ATmega8L)
  - 4.5V - 5.5V (ATmega8)
- Speed Grades
  - 0 - 8MHz (ATmega8L)
  - 0 - 16MHz (ATmega8)
- Power Consumption at 4Mhz, 3V, 25°C
  - Active: 3.6mA
  - Idle Mode: 1.0mA
  - Power-down Mode: 0.5µA

Atmel®

8-bit Atmel with 8KBytes In-System Programmable Flash

ATmega8
ATmega8L

Rev.2486AA–AVR–02/2013

Pin Configurations

PDIP

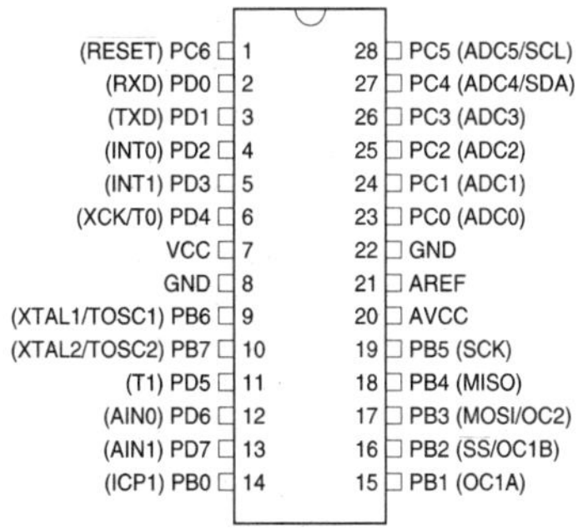

TQFP Top View

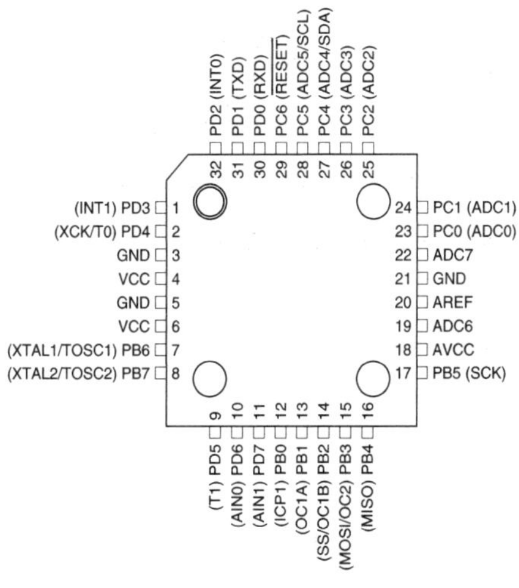

MLF Top View

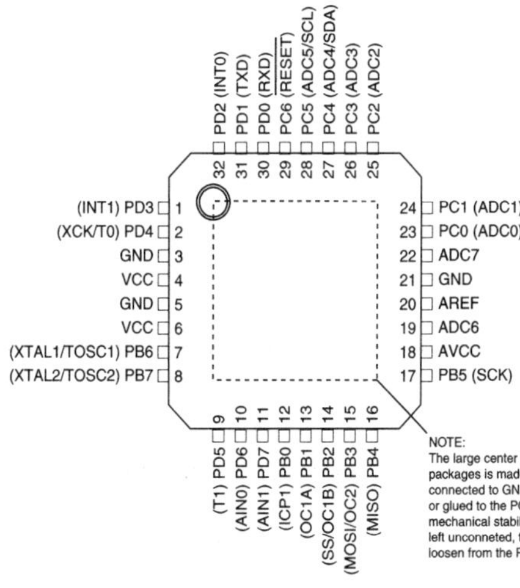

NOTE:
The large center pad underneath the MLF packages is made of metal and internally connected to GND. It should be soldered or glued to the PCB to ensure good mechanical stability. If the center pad is left unconneted, the package might loosen from the PCB.

Ordering Information

Speed (MHz)	Power Supply (V)	Ordering Code[2]	Package[1]	Operation Range
8	2.7 - 5.5	ATmega8L-8AU ATmega8L-8AUR[3] ATmega8L-8PU ATmega8L-8MU ATmega8L-8MUR[3]	32A 32A 28P3 32M1-A 32M1-A	Industrial (-40°C to 85°C)
16	4.5 - 5.5	ATmega8-16AU ATmega8-16AUR[3] ATmega8-16PU ATmega8-16MU ATmega8-16MUR[3]	32A 32A 28P3 32M1-A 32M1-A	
8	2.7 - 5.5	ATmega8L-8AN ATmega8L-8ANR[3] ATmega8L-8PN ATmega8L-8MN ATmega8L-8MUR[3]	32A 32A 28P3 32M1-A 32M1-A	Industrial (-40°C to 105°C)
16	4.5 - 5.5	ATmega8-16AN ATmega8-16ANR[3] ATmega8-16PN ATmega8-16MN ATmega8-16MUR[3]	32A 32A 28P3 32M1-A 32M1-A	

Notes:
1. This device can also be supplied in wafer form. Please contact your local Atmel sales office for detailed ordering information and minimum quantities
2. Pb-free packaging complies to the European Directive for Restriction of Hazardous Substances (RoHS directive). Also Halide free and fully Green
3. Tape & Reel
4. See characterization specification at 105°C

Package Type	
32A	32-lead, Thin (1.0mm) Plastic Quad Flat Package (TQFP)
28P3	28-lead, 0.300" Wide, Plastic Dual Inline Package (PDIP)
32M1-A	32-pad, 5 x 5 x 1.0 body, Lead Pitch 0.50mm Quad Flat No-Lead/Micro Lead Frame Package (QFN/MLF)

Packaging Information

32A

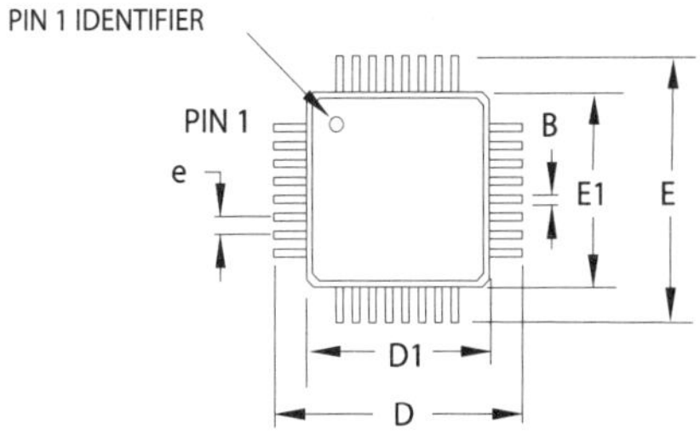

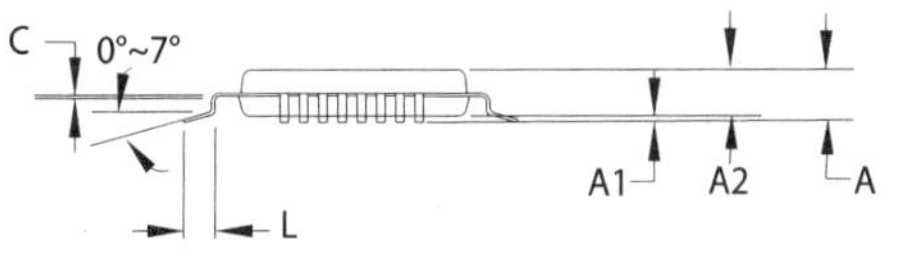

COMMON DIMENSIONS
(Unit of measure = mm)

SYMBOL	MIN	NOM	MAX	NOTE
A	–	–	1.20	
A1	0.05	–	0.15	
A2	0.95	1.00	1.05	
D	8.75	9.00	9.25	
D1	6.90	7.00	7.10	Note 2
E	8.75	9.00	9.25	
E1	6.90	7.00	7.10	Note 2
B	0.30	–	0.45	
C	0.09	–	0.20	
L	0.45	–	0.75	
e	0.80 TYP			

Notes:

1. This package conforms to JEDEC reference MS-026, Variation ABA.
2. Dimensions D1 and E1 do not include mold protrusion. Allowable protrusion is 0.25mm per side. Dimensions D1 and E1 are maximum plastic body size dimensions including mold mismatch.
3. Lead coplanarity is 0.10mm maximum.

2010-10-20

Atmel	TITLE 32A, 32-lead, 7 x 7mm body size, 1.0mm body thickness, 0.8mm lead pitch, thin profile plastic quad flat package (TQFP)	DRAWING NO. 32A	REV. C

28P3

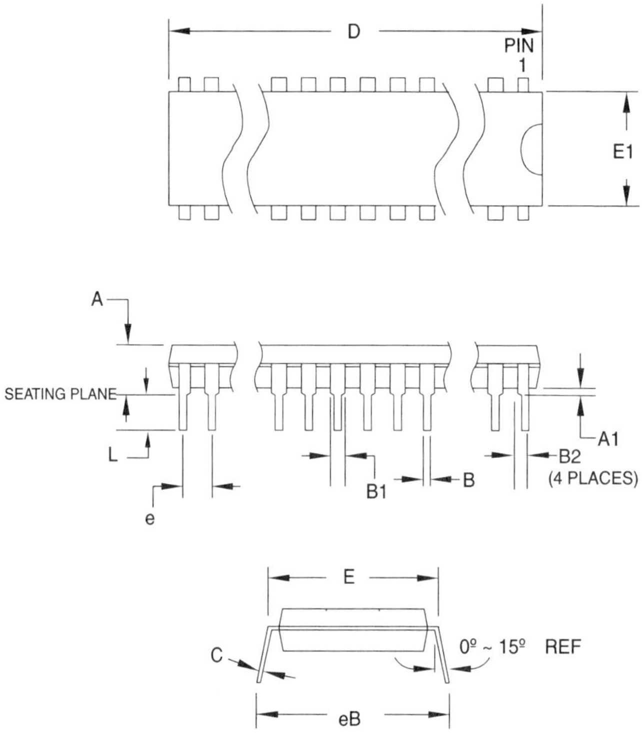

COMMON DIMENSIONS
(Unit of Measure = mm)

SYMBOL	MIN	NOM	MAX	NOTE
A	–	–	4.5724	
A1	0.508	–	–	
D	34.544	–	34.798	Note 1
E	7.620	–	8.255	
E1	7.112	–	7.493	Note 1
B	0.381	–	0.533	
B1	1.143	–	1.397	
B2	0.762	–	1.143	
L	3.175	–	3.429	
C	0.203	–	0.356	
eB	–	–	10.160	
e	2.540 TYP			

Note: 1. Dimensions D and E1 do not include mold Flash or Protrusion. Mold Flash or Protrusion shall not exceed 0.25mm (0.010").

09/28/01

ATMEL® 2325 Orchard Parkway San Jose, CA 95131	TITLE **28P3**, 28-lead (0.300"/7.62mm Wide) Plastic Dual Inline Package (PDIP)	DRAWING NO. 28P3	REV. B

32M1-A

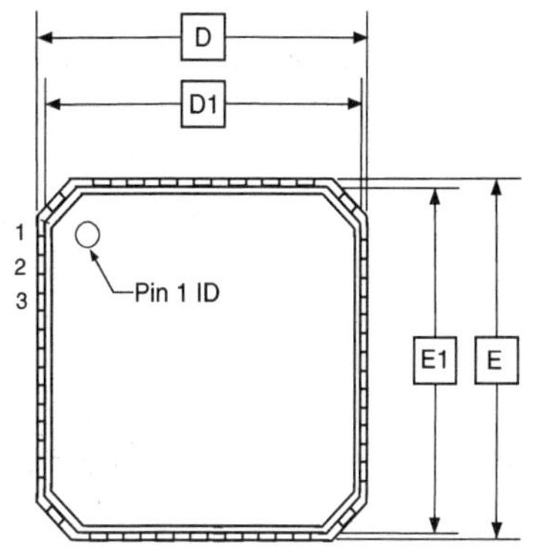

TOP VIEW

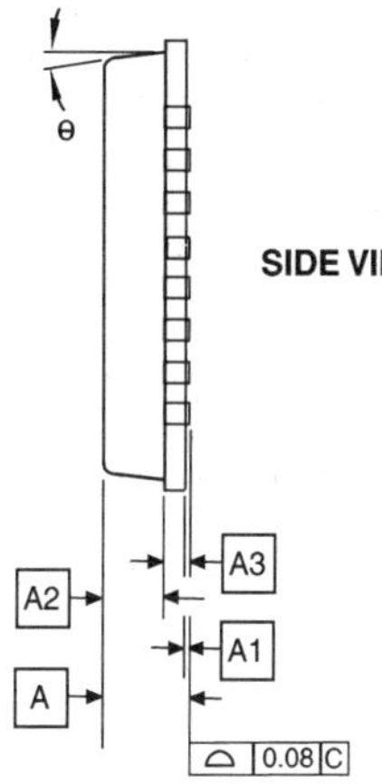

SIDE VIEW

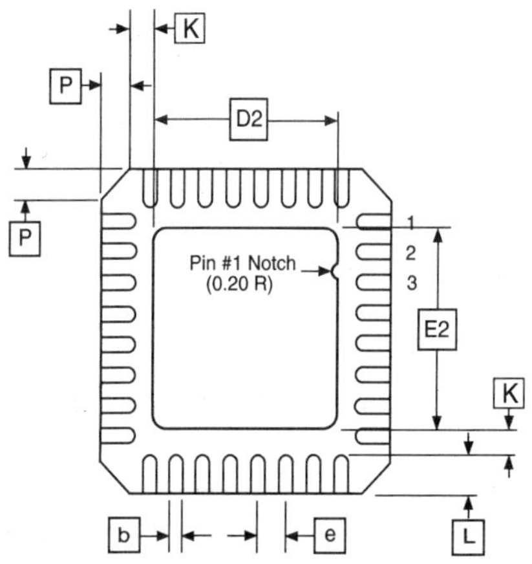

BOTTOM VIEW

COMMON DIMENSIONS
(Unit of Measure = mm)

SYMBOL	MIN	NOM	MAX	NOTE
A	0.80	0.90	1.00	
A1	–	0.02	0.05	
A2	–	0.65	1.00	
A3		0.20 REF		
b	0.18	0.23	0.30	
D	4.90	5.00	5.10	
D1	4.70	4.75	4.80	
D2	2.95	3.10	3.25	
E	4.90	5.00	5.10	
E1	4.70	4.75	4.80	
E2	2.95	3.10	3.25	
e		0.50 BSC		
L	0.30	0.40	0.50	
P	–	–	0.60	
θ	–	–	12^{0}	
K	0.20	–	–	

Note: JEDEC Standard MO-220, Fig. 2 (Anvil Singulation), VHHD-2.

5/25/06

ATMEL 2325 Orchard Parkway San Jose, CA 95131	TITLE **32M1-A**, 32-pad, 5 x 5 x 1.0mm Body, Lead Pitch 0.50mm, 3.10mm Exposed Pad, Micro Lead Frame Package (MLF)	DRAWING NO. 32M1-A	REV. E

MOTOROLA

LM358, LM258, LM2904, LM2904V

Dual Low Power Operational Amplifiers

Utilizing the circuit designs perfected for recently introduced uad Operational Amplifiers, these dual operational amplifiers feature 1) low power drain, 2) a common mode input voltage range extending to ground/V_{EE}, 3) single supply or split supply operation and 4) pinouts compatible with the popular C1558 dual operational amplifier. The L 158 series is e uivalent to one half of an L 124.

These amplifiers have several distinct advantages over standard operational amplifier types in single supply applications. They can operate at supply voltages as low as 3.0 V or as high as 32 V, with uiescent currents about one fifth of those associated with the C1741 (on a per amplifier basis). The common mode input range includes the negative supply, thereby eliminating the necessity for external biasing components in many applications. The output voltage range also includes the negative power supply voltage.

- Short Circuit Protected Outputs
- True Differential Input Stage
- Single Supply Operation: 3.0 V to 32 V
- Low Input Bias Currents
- Internally Compensated
- Common ode Range Extends to Negative Supply
- Single and Split Supply Operation
- Similar Performance to the Popular C1558
- ESD Clamps on the Inputs Increase Ruggedness of the Device without Affecting Operation

DUAL DIFFERENTIAL INPUT OPERATIONAL AMPLIFIERS

SEMICONDUCTOR TECHNICAL DATA

N SUFFIX
PLASTIC PAC AGE
CASE 626

D SUFFIX
PLASTIC PAC AGE
CASE 751
(SO 8)

PIN CONNECTIONS

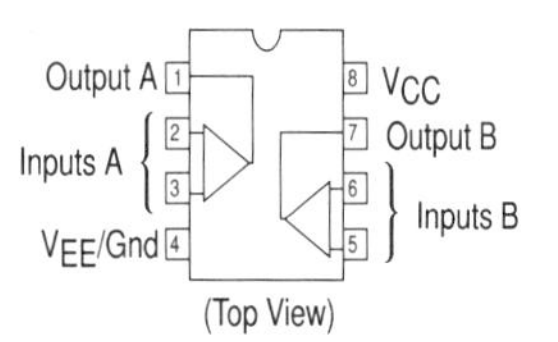

MAXIMUM RATINGS (T_A 25 °C, unless otherwise noted.)

Rating	Symbol	LM258 LM358	LM2904 LM2904V	Unit
Power Supply Voltages Single Supply Split Supplies	 V_{CC} V_{CC}, V_{EE}	 32 ±16	 26 ±13	Vdc
Input Differential Voltage Range (Note 1)	V_{IDR}	±32	±26	Vdc
Input Common ode Voltage Range (Note 2)	V_{ICR}	0.3 to 32	0.3 to 26	Vdc
Output Short Circuit Duration	t_{SC}	Continuous		
Junction Temperature	T_J	150		°C
Storage Temperature Range	T_{stg}	55 to 125		°C
Operating Ambient Temperature Range L 258 L 358 L 2904 L 2904V	T_A	 25 to 85 0 to 70	 40 to 105 40 to 125	°C

NOTES: 1. Split Power Supplies.
2. For Supply Voltages less than 32 V for the L 258/358 and 26 V for the L 2904, the absolute maximum input voltage is e ual to the supply voltage.

ORDERING INFORMATION

Device	Operating Temperature Range	Package
L 2904D	T_A 40 ° to 105 °C	SO 8
L 2904N		Plastic DIP
L 2904VD	T_A 40 ° to 125 °C	SO 8
L 2904VN		Plastic DIP
L 258D	T_A 25 ° to 85 °C	SO 8
L 258N		Plastic DIP
L 358D	T_A 0 ° to 70 °C	SO 8
L 358N		Plastic DIP

OUTLINE DIMENSIONS

N SUFFIX
PLASTIC PAC AGE
CASE 626 05
ISSUE

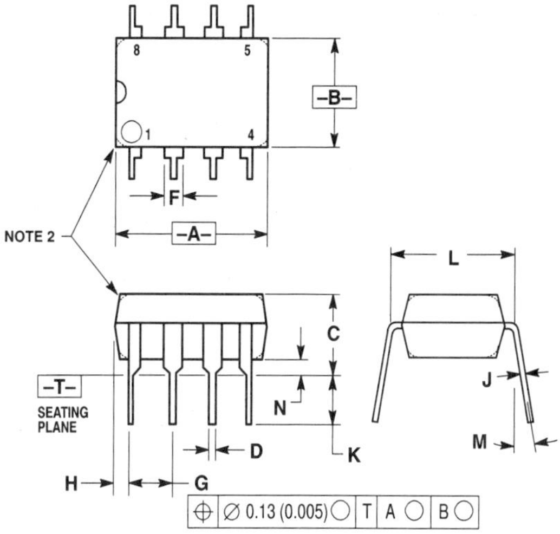

NOTES:
1. DI ENSION L TO CENTER OF LEAD WHEN FOR ED PARALLEL.
2. PAC AGE CONTOUR OPTIONAL (ROUND OR S UARE CORNERS).
3. DI ENSIONING AND TOLERANCING PER ANSI 14.5 , 1982.

DIM	MILLIMETERS MIN	MILLIMETERS MAX	INCHES MIN	INCHES MAX
A	9.40	10.16	0.370	0.400
B	6.10	6.60	0.240	0.260
C	3.94	4.45	0.155	0.175
D	0.38	0.51	0.015	0.020
F	1.02	1.78	0.040	0.070
G	2.54 BSC		0.100 BSC	
H	0.76	1.27	0.030	0.050
J	0.20	0.30	0.008	0.012
K	2.92	3.43	0.115	0.135
L	7.62 BSC		0.300 BSC	
M		10 °		10 °
N	0.76	1.01	0.030	0.040

D SUFFIX
PLASTIC PAC AGE
CASE 751 05
(SO 8)
ISSUE R

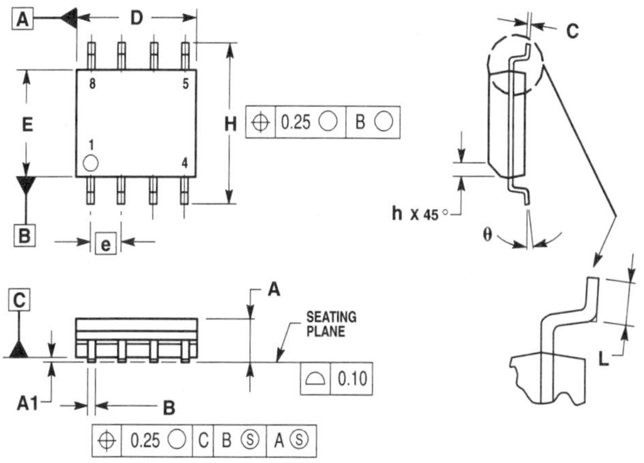

NOTES:
1. DI ENSIONING AND TOLERANCING PER AS E 14.5 , 1994.
2. DI ENSIONS ARE IN ILLI ETERS.
3. DI ENSION D AND E DO NOT INCLUDE OLD PROTRUSION.
4. AXI U OLD PROTRUSION 0.15 PER SIDE.
5. DI ENSION B DOES NOT INCLUDE OLD PROTRUSION. ALLOWABLE DA BAR PROTRUSION SHALL BE 0.127 TOTAL IN EXCESS OF THE B DI ENSION AT AXI U ATERIAL CONDITION.

DIM	MILLIMETERS MIN	MILLIMETERS MAX
A	1.35	1.75
A1	0.10	0.25
B	0.35	0.49
C	0.18	0.25
D	4.80	5.00
E	3.80	4.00
e	1.27 BSC	
H	5.80	6.20
h	0.25	0.50
L	0.40	1.25
θ	0°	7°

MBS/YJ SERIES

SINGLE PHASE 0.5 AMPS.
GLASS PASSIVATED BRIDGE
RECTIFIERS

Voltage Range
200 to 1000 Volts
Current
0.5 Amperes

Features

- Ideal for printed circuit board
  Reliable low cost construction utilizing molded plastic technique
- High surge current capability
- High temperature soldering guaranteed: 260°C / 10 seconds at 5 lbs.,(2.3kg) tension
- Small size,simple installation
- Leads solderable per MIL-STD-202 Method 208

MBS

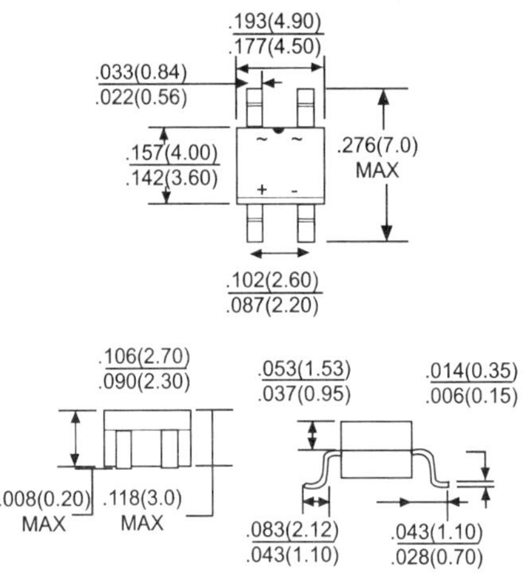

Dimensions in inches and (millimeters)

MAXIMUM RATINGS AND ELECTRICALCHARACTERISTICS

Rating at 25°C ambient temperature unless otherwise specified.
Single phase, half wave, 60Hz, resistive or inductive load.
For capacitive load, derate current by 20%

Type Number		MB2S YJ2J	MB4S YJ4J	MB6S YJ5J	MB8S YJ6J	MB10S YJ7J	UNITS
Maximum Repetitive Peak Reverse Voltage	V_{RRM}	200	400	600	800	1000	V
Maximum RMS Voltage	V_{RMS}	140	280	420	560	700	V
Maximum DC Blocking Voltage	V_{DC}	200	400	600	800	1000	V
Maximunm Average Forward Rectified Current at T_A = 30°C On glass-epoxy P.C.B. / On aluminum substrate	$I_{F(AV)}$	0.5 0.8					A
Peak Forward Surge Current, 8.3 ms Single Half Sinewave Superimposed on Rated Load (JEDEC method)	I_{FSM}	35					A
Maximum Instantaneous Forward Voltage Drop Per Leg@0.4A	V_F	1.0					V
Maximun DC Reverse Current T_A = 25°C at Rated DC Blocking Voltage T_A = 100°C	I_R	5.0 100					uA uA
Typical Junction Capacitance Per Leg (Note3)	C_J	15					pF
Typical Thermal Resistance Per Leg	$R_{\theta JA}$	75					°C/W
Operating Temperature Range	T_J	-55 to +150					°C
Storage Temperature Range	T_{STG}	-55 to +150					°C

Note:1.On glass epoxy P.C.B mounted on 0.05x 0.05"(1.3x1.3mm)pads
2.On aluminum substrate P.C.B with on area of 0.8"x 0.8"(20x20mm) mounted on 0.05x 0.05"(1.3x1.3mm)solder pad
3.Measured at 1.0MHz and applied reverse valtage of 4.0 volts.

6-Lead Thin Shrink Small Outline Transistor Package [SC70] (KS-6)

Dimensions shown in millimeters

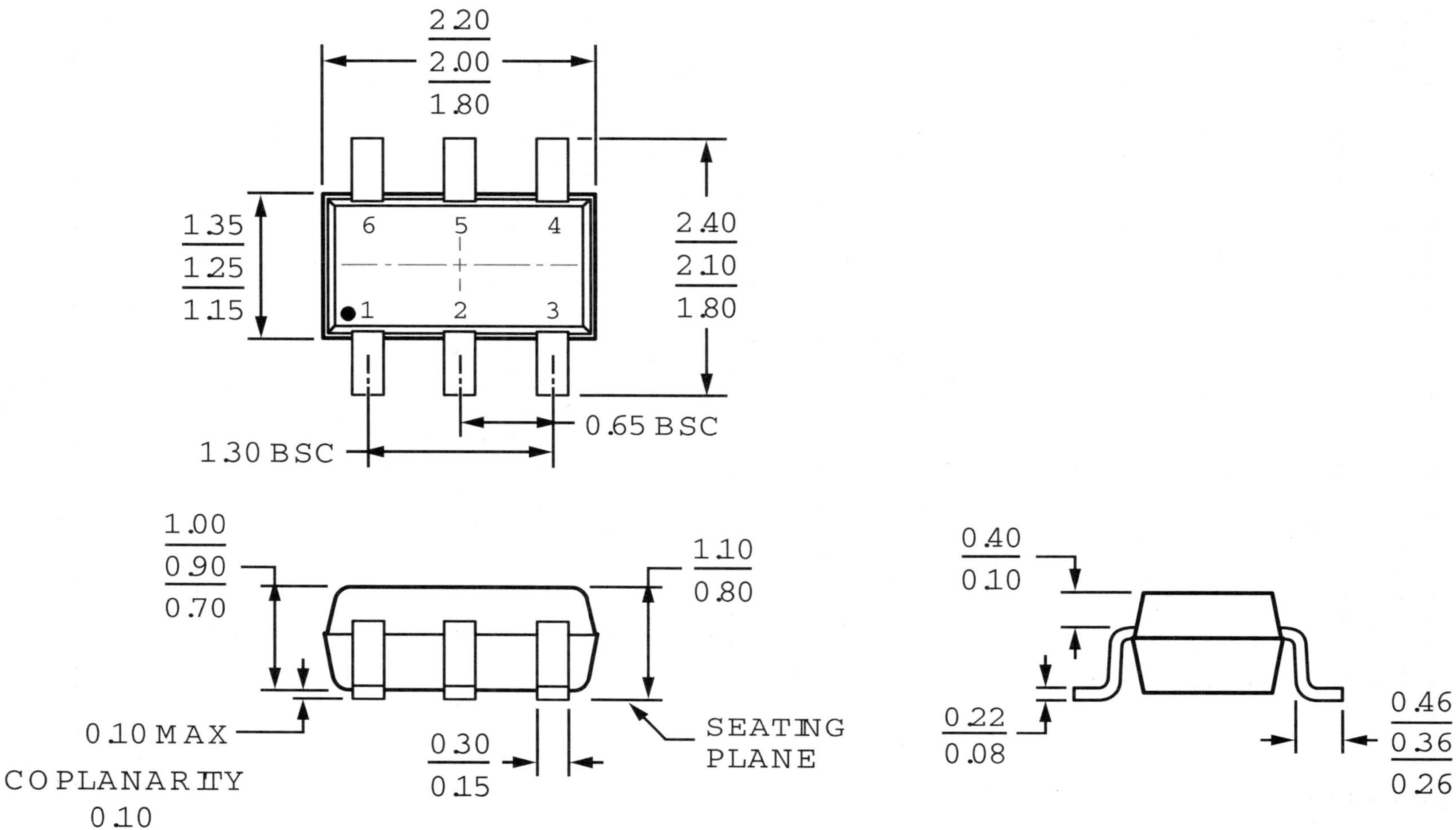

COMPLIANT TO JEDEC STANDARDS MO-203-AB

072809-A

- 3-Terminal Regulators
- Output Current up to 1.5 A
- Internal Thermal-Overload Protection
- High Power-Dissipation Capability
- Internal Short-Circuit Current Limiting
- Output Transistor Safe-Area Compensation

KC (TO-220) PACKAGE (TOP VIEW)

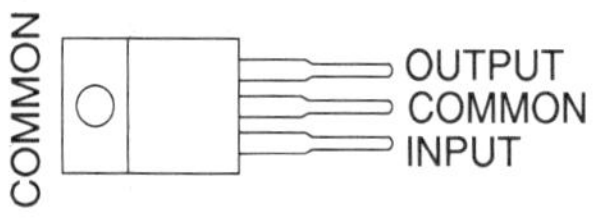

KCS (TO-220) PACKAGE (TOP VIEW)

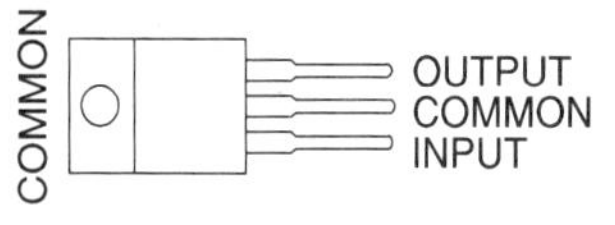

KTE PACKAGE (TOP VIEW)

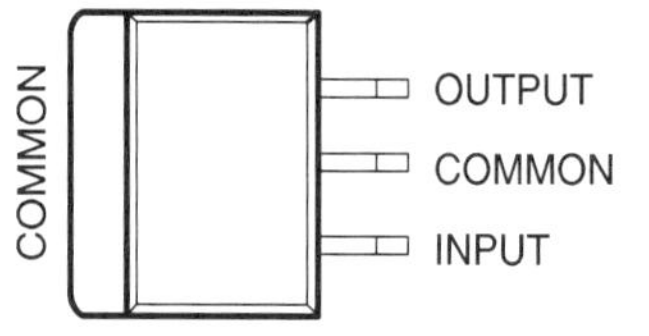

description/ordering information

This series of fixed-voltage integrated-circuit voltage regulators is designed for a wide range of applications. These applications include on-card regulation for elimination of noise and distribution problems associated with single-point regulation. Each of these regulators can deliver up to 1.5 A of output current. The internal current-limiting and thermal-shutdown features of these regulators essentially make them immune to overload. In addition to use as fixed-voltage regulators, these devices can be used with external components to obtain adjustable output voltages and currents, and also can be used as the power-pass element in precision regulators.

ORDERING INFORMATION

T_J	$V_{O(NOM)}$ (V)	PACKAGE†		ORDERABLE PART NUMBER	TOP-SIDE MARKING
0°C to 125°C	5	POWER-FLEX (KTE)	Reel of 2000	µA7805CKTER	µA7805C
		TO-220 (KC)	Tube of 50	µA7805CKC	µA7805C
		TO-220, short shoulder (KCS)	Tube of 20	µA7805CKCS	
	8	POWER-FLEX (KTE)	Reel of 2000	µA7808CKTER	µA7808C
		TO-220 (KC)	Tube of 50	µA7808CKC	µA7808C
		TO-220, short shoulder (KCS)	Tube of 20	µA7808CKCS	
	10	POWER-FLEX (KTE)	Reel of 2000	µA7810CKTER	µA7810C
		TO-220 (KC)	Tube of 50	µA7810CKC	µA7810C
	12	POWER-FLEX (KTE)	Reel of 2000	µA7812CKTER	µA7812C
		TO-220 (KC)	Tube of 50	µA7812CKC	µA7812C
		TO-220, short shoulder (KCS)	Tube of 20	µA7812CKCS	
	15	POWER-FLEX (KTE)	Reel of 2000	µA7815CKTER	µA7815C
		TO-220 (KC)	Tube of 50	µA7815CKC	µA7815C
		TO-220, short shoulder (KCS)	Tube of 20	µA7815CKCS	
	24	POWER-FLEX (KTE)	Reel of 2000	µA7824CKTER	µA7824C
		TO-220 (KC)	Tube of 50	µA7824CKC	µA7824C

† Package drawings, standard packing quantities, thermal data, symbolization, and PCB design guidelines are available at www.ti.com/sc/package.

Please be aware that an important notice concerning availability, standard warranty, and use in critical applications of Texas Instruments semiconductor products and disclaimers thereto appears at the end of this data sheet.

KTE (R-PSFM-G3) **PowerFLEX™ PLASTIC FLANGE-MOUNT**

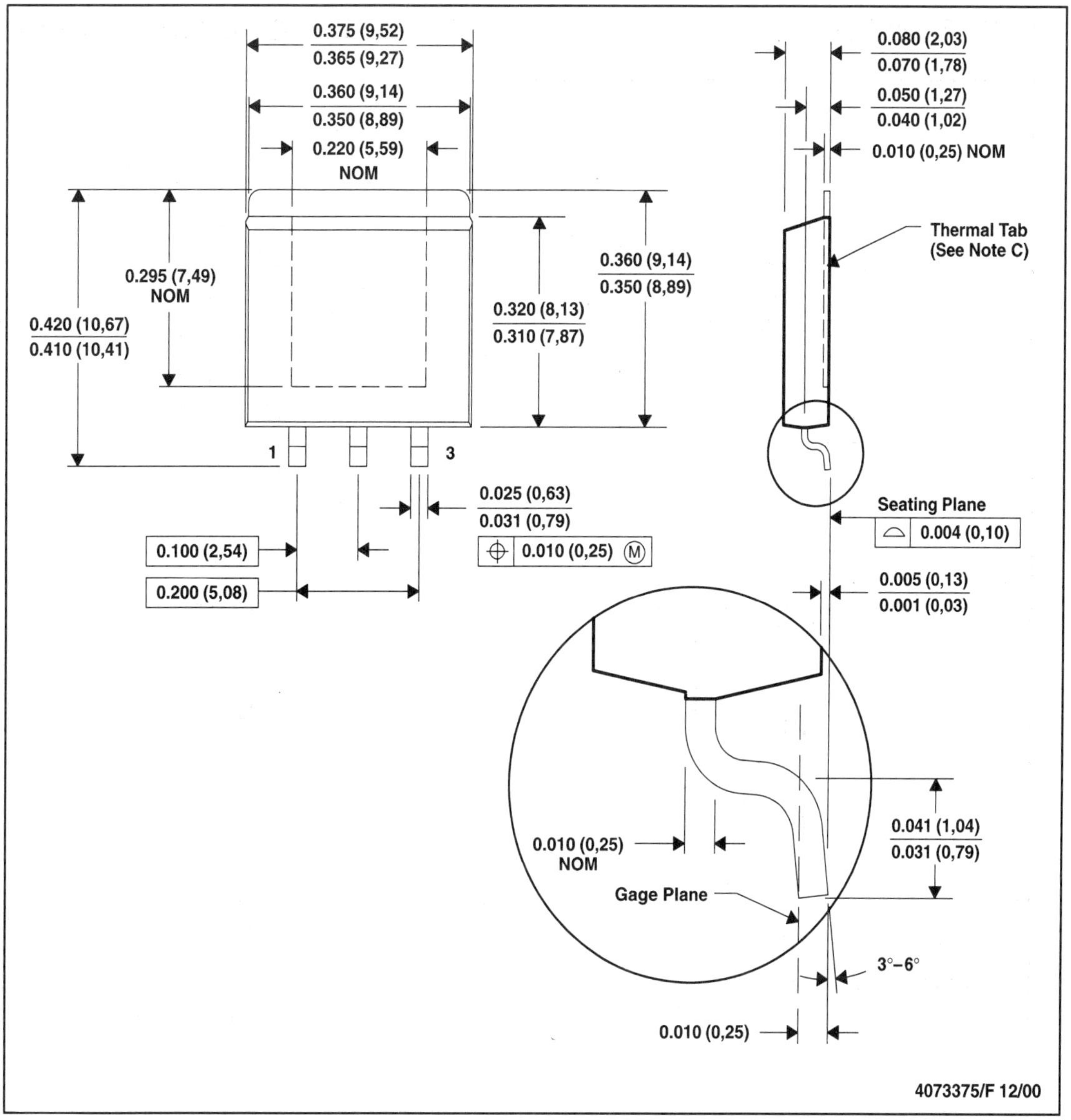

NOTES: A. All linear dimensions are in inches (millimeters).
B. This drawing is subject to change without notice.
C. The center lead is in electrical contact with the thermal tab.
D. Dimensions do not include mold protrusions, not to exceed 0.006 (0,15).
E. Falls within JEDEC MO-169

PowerFLEX is a trademark of Texas Instruments.

KCS (R–PSFM–T3) PLASTIC FLANGE–MOUNT PACKAGE

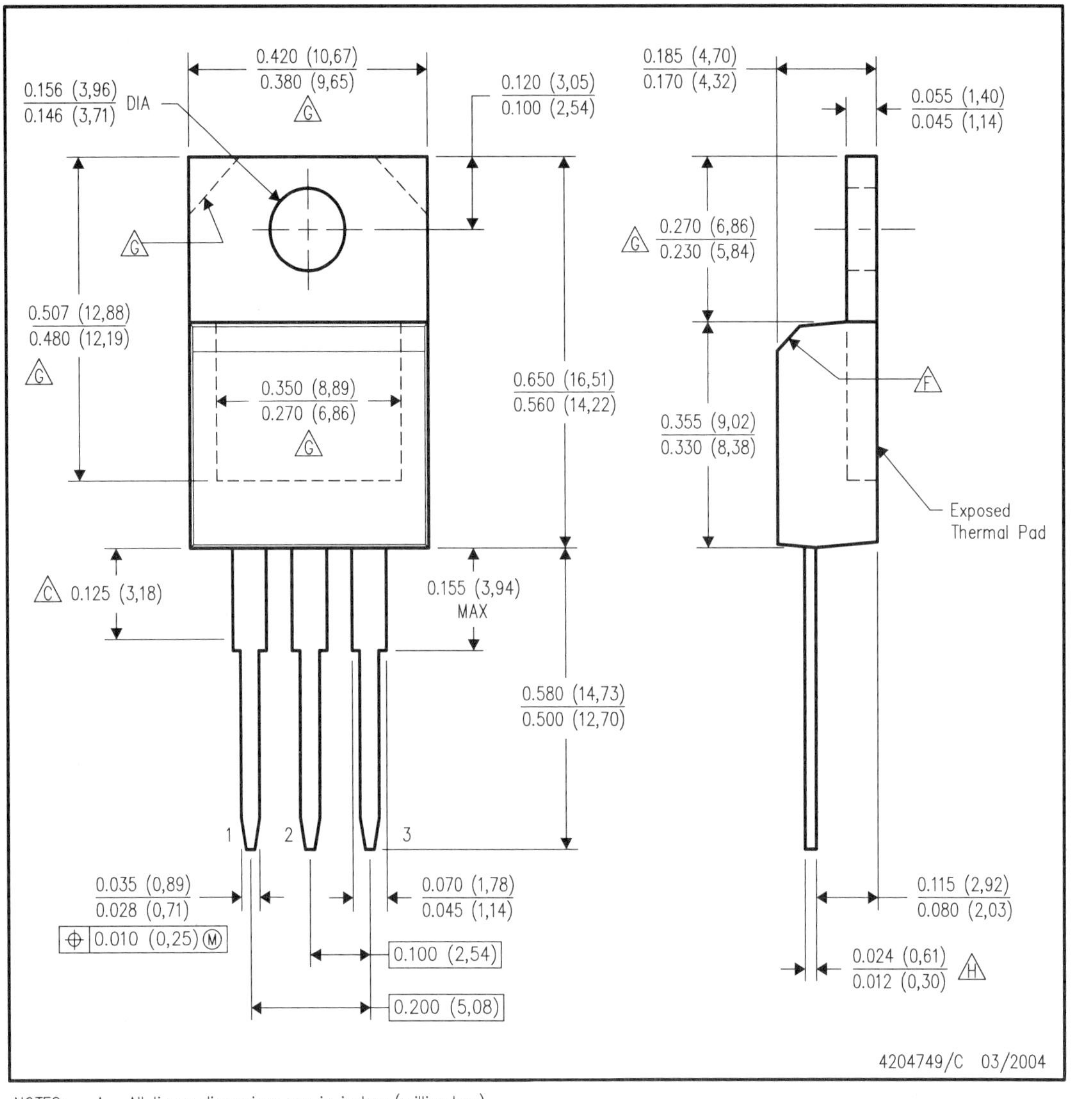

NOTES:
A. All linear dimensions are in inches (millimeters).
B. This drawing is subject to change without notice.
C. Lead dimensions are not controlled within this area.
D. All lead dimensions apply before solder dip.
E. The center lead is in electrical contact with the mounting tab.
F. The chamfer is optional.
G. Thermal pad contour optional within these dimensions.
H. Falls within JEDEC TO–220 variation AB, except minimum lead thickness.

KC (R–PSFM–T3) PLASTIC FLANGE–MOUNT PACKAGE

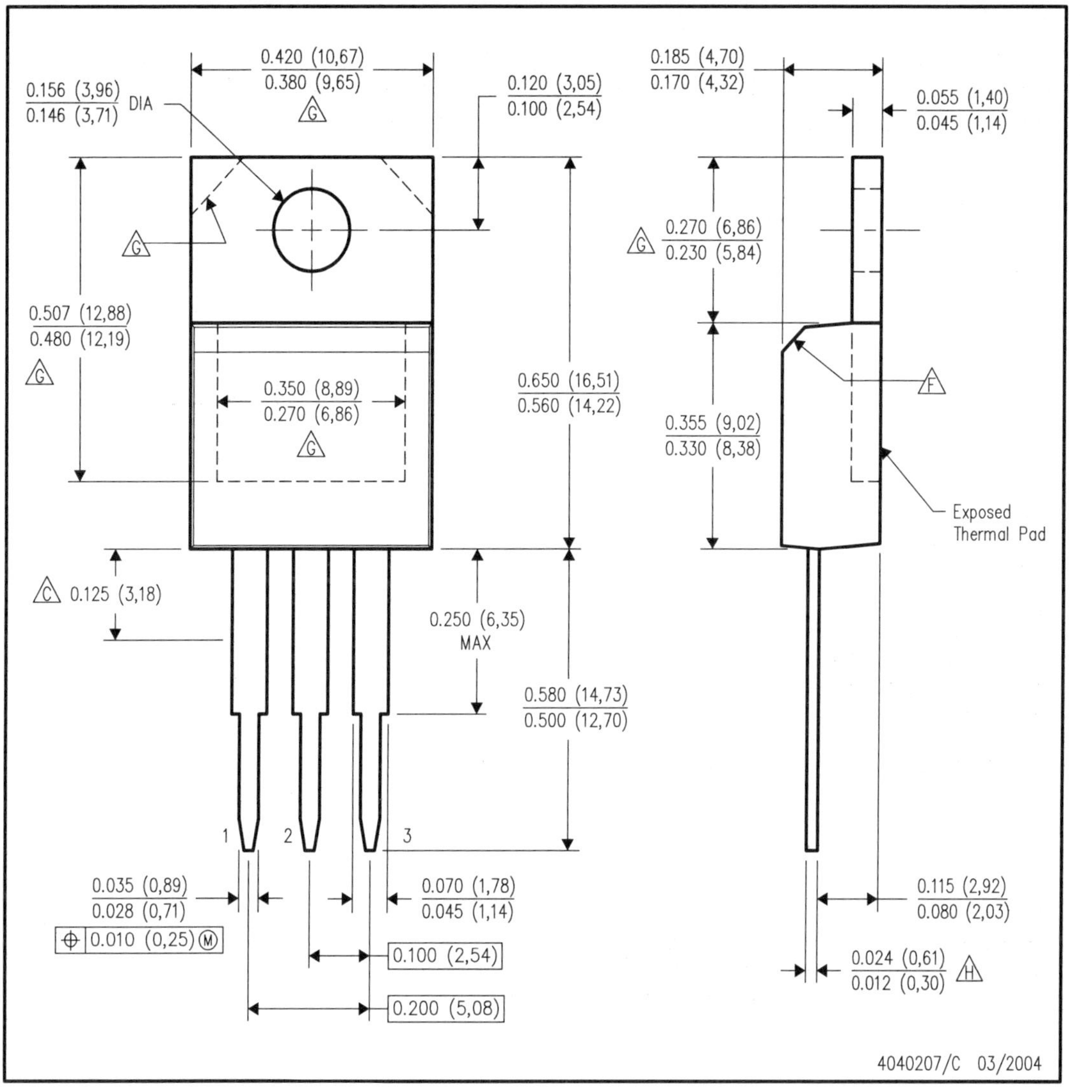

NOTES:
- A. All linear dimensions are in inches (millimeters).
- B. This drawing is subject to change without notice.
- C. Lead dimensions are not controlled within this area.
- D. All lead dimensions apply before solder dip.
- E. The center lead is in electrical contact with the mounting tab.
- F. The chamfer is optional.
- G. Thermal pad contour optional within these dimensions.
- H. Falls within JEDEC TO–220 variation AB, except minimum lead thickness.

EVERLIGHT

DATASHEET

5mm Infrared LED,T-1 3/4
IR383

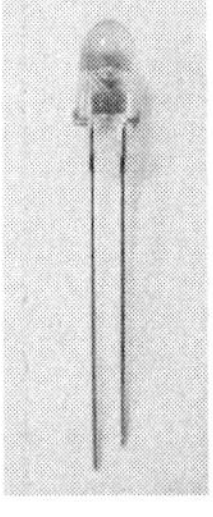

Features

- High reliability
- High radiant intensity
- Peak wavelength λp=940nm
- 2.54mm Lead spacing
- Low forward voltage
- Pb Free
- This product itself will remain within RoHS compliant version.

Description

- EVERLIGHT's Infrared Emitting Diode (IR383) is a high intensity diode , molded in a blue transparent plastic package.
- The device is spectrally matched with phototransistor , photodiode and infrared receiver module.

Applications

- Infrared applied system

Package Dimension

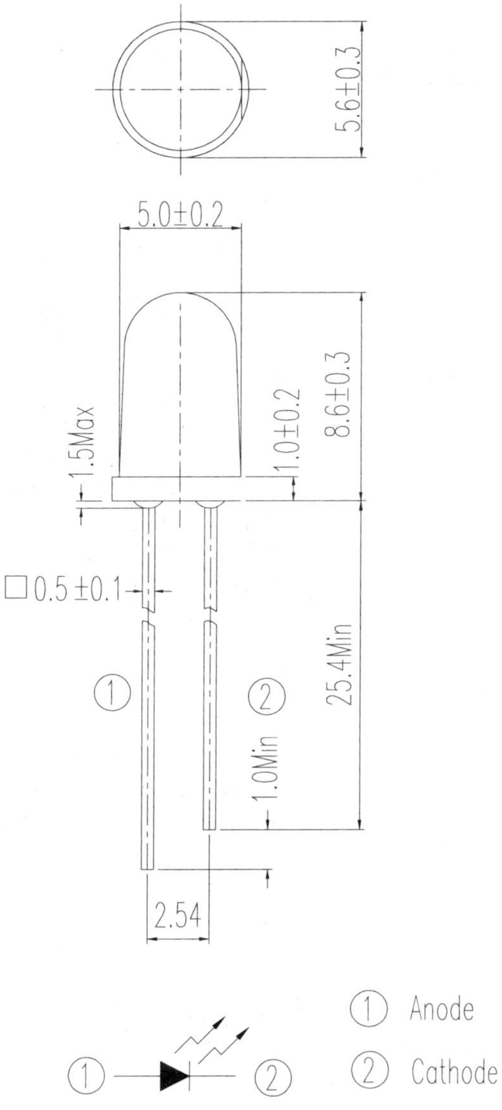

Note: Tolerances unless dimensions ±0.25mm

HALF-SIZE DIP LOW VOLTAGE 5.0V CRYSTAL CLOCK OSCILLATOR

ACH Series

RoHS Compliant

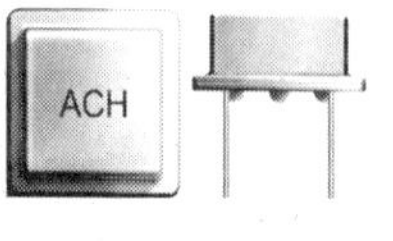

12.7 x 12.7 x 5.6mm

FEATURES:

- Wide frequency range.
- HCMOS and TTL compatible.
- Tight symmetry option.
- Tri-state Enable / Disable function.

APPLICATIONS:

- Microprocessors.
- Digital circuit.

STANDARD SPECIFICATIONS:

PARAMETERS	
ABRACON P/N	ACH Series
Frequency:	32.768kHz to 200MHz
Operating temperature:	0°C to +70°C
Storage temperature:	- 55°C to + 125°C
Frequency Stability:	± 100ppm max. (see options)
Supply voltage (Vdd):	5.0Vdc ± 10%
Symmetry:	45/55% max.for F<50MHz 40/60% max. for F>=50MHz (see options)
Rise and Fall Times(Tr/Tf):	40ns max. for 32.768kHz ≤ F < 2MHz 10ns max. for 2MHz ≤ F ≤ 24MHz 5ns max. for F>24MHz
Output load:	10 TTL or 15 pF for F<80MHz 5 TTL or 15 pF for F>=80MHz
Output Voltage:	VOH = 0.9 * VDD min. VOL = 0.1*VDD max.
Tri-state Function:	"1" (VIH >= 2.2 Vdc) or open: Oscillation "0" (VIL < 0.8V): Hi Z
Start-up-time:	10 ms max
Output Disable/ Enable time:	100 ns max. (for Option "-A" ONLY)
Input current:	30 mA max. for F < 24 MHz 45 mA max. for F < 50 MHz 70 mA max. for F < 80 MHz 85 mA max. for F >= 24 MHz

OPTIONS & PART IDENTIFICATION:

(Left blank if standard)

ACH - Frequency - □ - □ - □ - □ - □ - □ - □

Temperature Options	
D	-10°C to +60°C
E	-20°C to +70°C
F	-30°C to +70°C
N	-30°C to +85°C
L	-40°C to +85°C

Freq. Stability	
Y	±10 ppmmax.
J	±20 ppm max.
R	±25 ppm max.
K	±30 ppm max.
H	±35 ppm max.
C	±50 ppm max.

Symmetry Options	
S	45/55%@1/2Vdd
T	47.5/52.5%@1/2Vdc
S1	45/55@1.4Vdc
T1	45.5/52.5@1.4Vdc

Tristate Enable Funct	
A	Tristate Function

50pF Option	
50	50pF

Packaging	
Blank	Tube
TY	Tray

Value Added Option	
G	Gull Wing
G2	Gull Wing
GHT	Gull Wing Reflow
Q30	0.30(7.62)
Q25	0.25(6.35)
Q20	0.20(5.08)
Q15	0.15(3.81)
Q10	0.10(2.54)

30332 Esperanza, Rancho Santa Margarita, California 92688
tel 949-546-8000 | fax 949-546-8001 | www.abracon.com

HALF-SIZE DIP LOW VOLTAGE 5.0V CRYSTAL CLOCK OSCILLATOR

ACH Series

RoHS Compliant

12.7 x 12.7 x 5.6mm

OUTLINE DIMENSIONS:

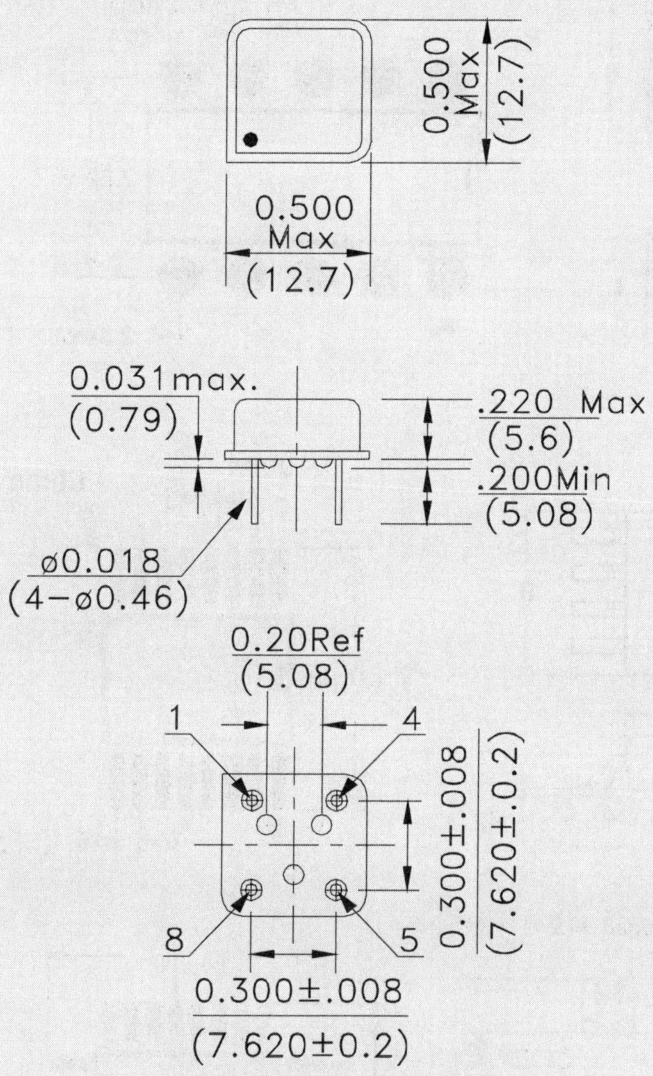

PIN	FUNCTION
1	NC or Tri-state Enable/
4	GND/Case
5	Output
8	VDD

Dimensions: inches (mm)

● Wizard를 이용한 PCB Decal 만들기

1) Dual 14pin - Dip

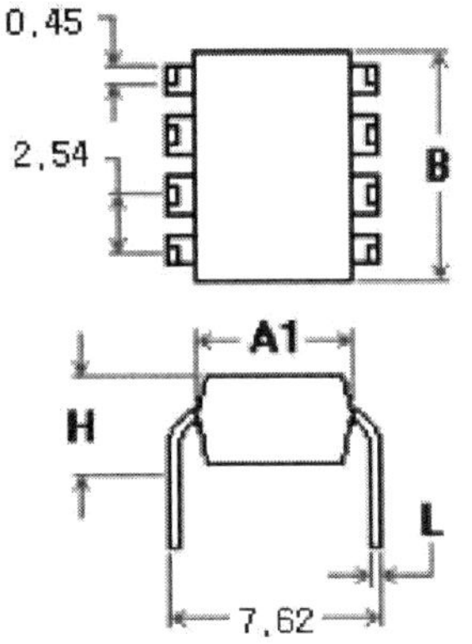

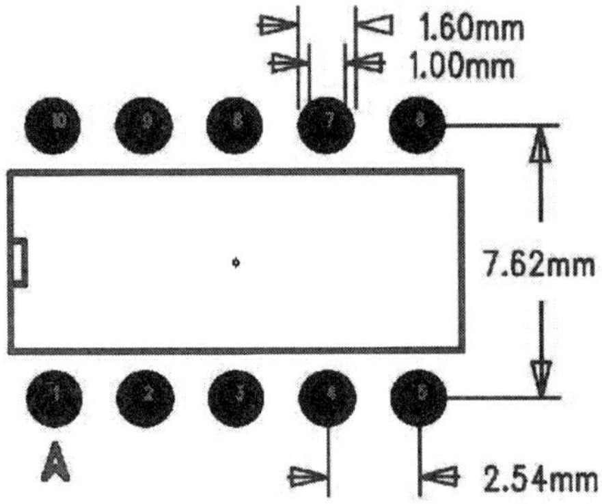

2) Dual 14pin - SMD

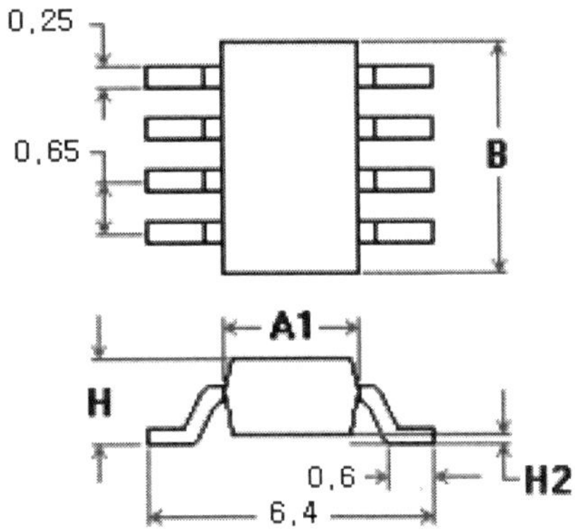

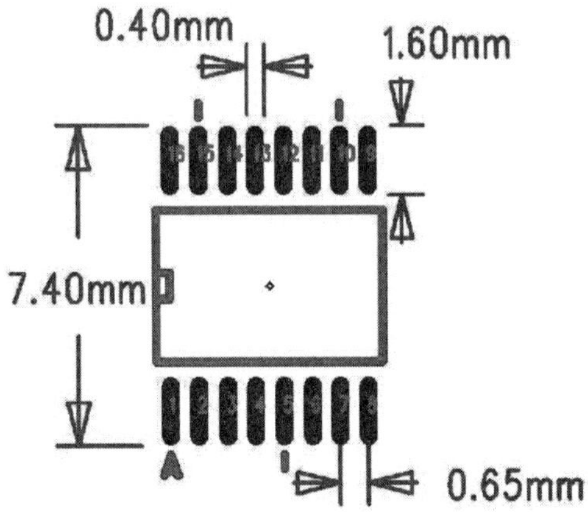

3) Quad 32pin (atmega8 회로에 사용)

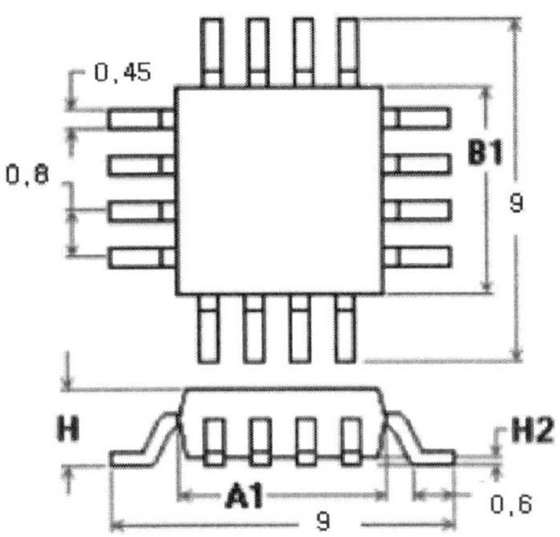

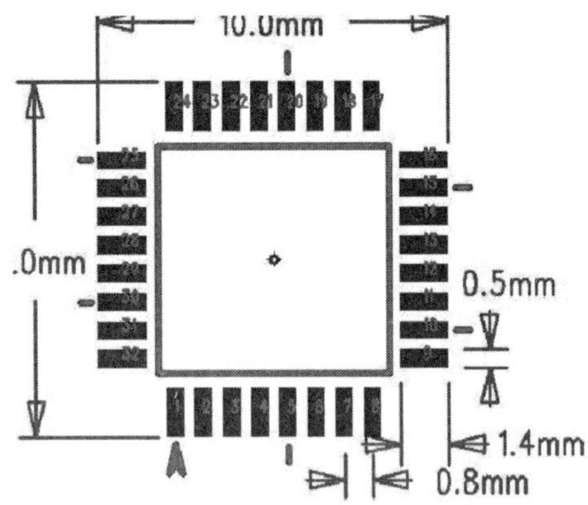

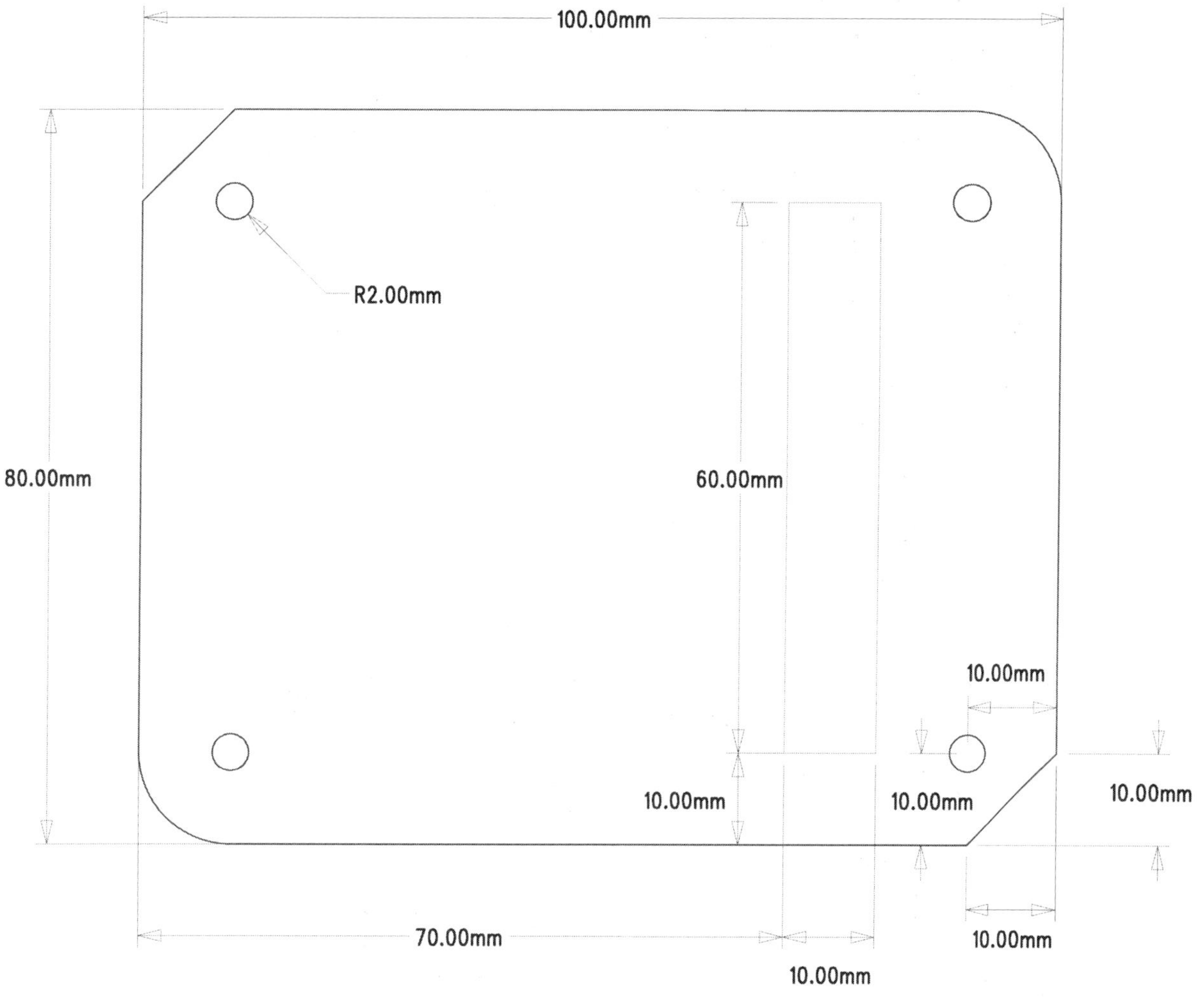
100.00mm
R2.00mm
80.00mm
60.00mm
10.00mm
10.00mm
10.00mm
10.00mm
70.00mm
10.00mm
10.00mm

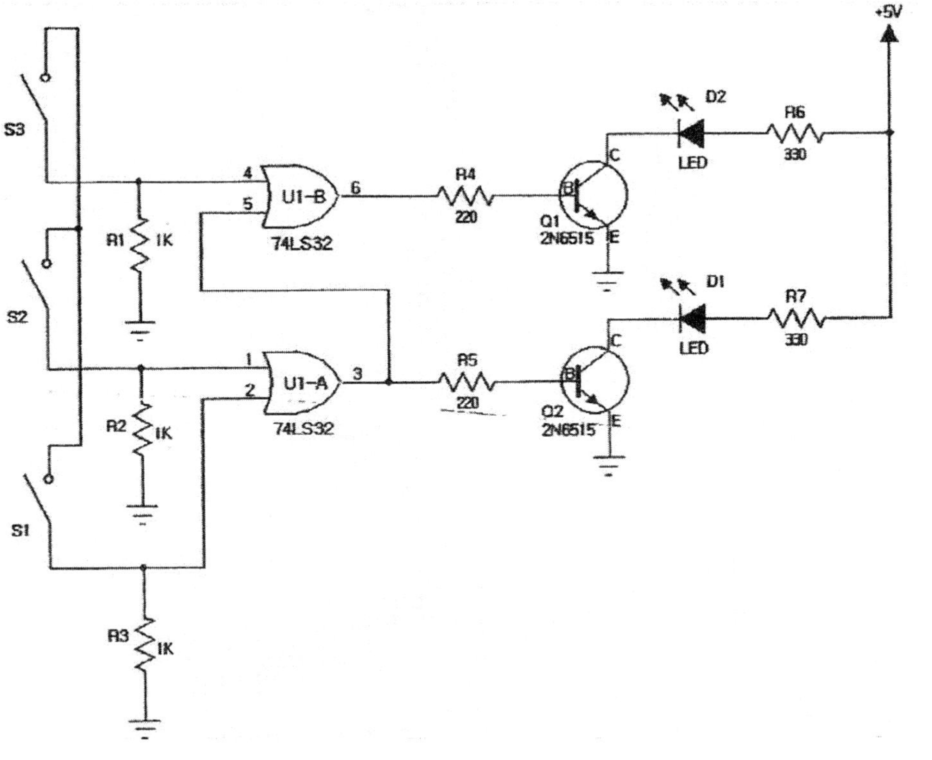

No.	수량	Ref-Des.	Part Type	PCB Decal	Value
1	2	Q1-2	2N6515	TO-220-UP	
2	1	U1	74LS32	DIP14	
3	2	D1-2	LED	LED	
4	3	S1-3	SW-NKK-JF15	SW	
5	3	R1-3	RES-1/8W	R1/8W	1K
6	2	R4-5	RES-1/8W	R1/8W	220
7	2	R6-7	RES-1/8W	R1/8W	330
Total	15				

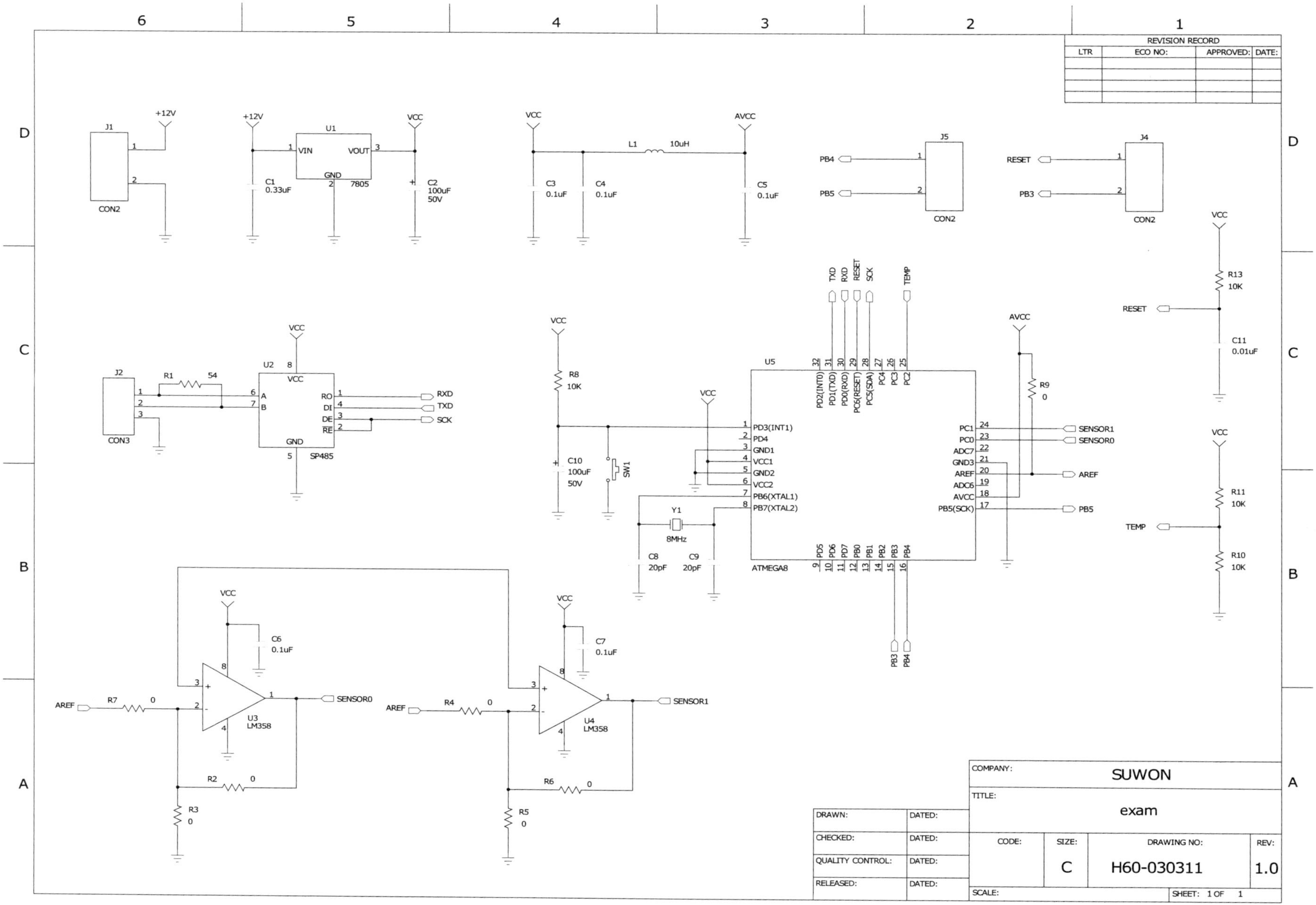
REVISION RECORD
LTR ECO NO: APPROVED: DATE:
J1 CON2 +12V
U1 7805 VIN VOUT GND
C1 0.33uF
C2 100uF 50V
L1 10uH
C3 0.1uF
C4 0.1uF
C5 0.1uF
J5 CON2 PB4 PB5
J4 CON2 RESET PB3
R13 10K
C11 0.01uF
R11 10K
R10 10K
TEMP
J2 CON3
R1 54
U2 SP485 VCC RO DI DE RE GND A B
RXD TXD SCK
R8 10K
C10 100uF 50V
SW1
Y1 8MHz
C8 20pF
C9 20pF
U5 ATMEGA8
PD3(INT1) PD4 GND1 VCC1 GND2 VCC2 PB6(XTAL1) PB7(XTAL2)
PD2(INT0) PD1(TXD) PD0(RXD) PC6(RESET) PC5(SDA) PC4 PC3 PC2
PC1 PC0 ADC7 GND3 AREF ADC6 AVCC PB5(SCK)
PD5 PD6 PD7 PB0 PB1 PB2 PB3 PB4
R9 0
SENSOR1 SENSOR0 AREF PB5
U3 LM358
U4 LM358
C6 0.1uF
C7 0.1uF
R7 0
R4 0
R2 0
R6 0
R3 0
R5 0
COMPANY: SUWON
TITLE: exam
DRAWN: DATED:
CHECKED: DATED:
QUALITY CONTROL: DATED:
RELEASED: DATED:
CODE: SIZE: C DRAWING NO: H60-030311 REV: 1.0
SCALE: SHEET: 1 OF 1

PADS를 이용한 PCB 설계 실무핵심

PCB 설계 실무

3판 2쇄 2025년 1월 20일

저 자 윤왕철 / 김규희 / 함영주
발행인 송광헌
발행처 도서출판 복두(더)
주 소 서울특별시 영등포구 경인로82길 3-4
센터플러스 807호
(우) 07371
전 화 02-2164-2580
FAX 02-2164-2584
등 록 1993. 11. 22. 제 10-902 호

정가 : 15,000원
ISBN : 979-11-6675-087-8 93560

한국과학기술출판협회 회원사